Smart Microgrid Systems

This book highlights microgrids as integrating platforms for distributed generation units, energy storages and local loads, with an emphasis on system performance via innovative approaches. It explains the smart power system concept, transmission, distribution, and utilization, and then looks at distributed generation technologies and hybrid power systems. Smart approaches, an analysis of microgrid design architecture and its implementation, the mitigation of cyber threats, and system optimization are also included. Case studies related to microgrid modeling and simulation are placed at the end of each chapter.

FEATURES

- Focuses on applications of expert systems for microgrid control
- Explores microgrid applications for power networks and applications of expert technologies
- Reviews design and development technologies related to renewable energy for a weak power network
- Discusses cyber security for microgrids
- Includes case studies related to actual developments and research

This book is aimed at researchers and graduate students in power engineering and electronics.

Smart Microgrid Systems
Advanced Technologies

KTM Udayanga Hemapala
MK Perera

CRC Press
Taylor & Francis Group
Boca Raton London New York

CRC Press is an imprint of the
Taylor & Francis Group, an **informa** business

First edition published 2023
by CRC Press
6000 Broken Sound Parkway NW, Suite 300, Boca Raton, FL 33487-2742

and by CRC Press
4 Park Square, Milton Park, Abingdon, Oxon, OX14 4RN

CRC Press is an imprint of Taylor & Francis Group, LLC

© 2023 KTM Udayanga Hemapala and MK Perera

ISBN: 978-1-032-10629-8 (hbk)
ISBN: 978-1-032-10630-4 (pbk)
ISBN: 978-1-003-21629-2 (ebk)

DOI: 10.1201/9781003216292

Typeset in Times
by Newgen Publishing UK

Contents

Preface...xi
Acknowledgements..xv
Authors..xvii
Introduction...xix

Chapter 1   Overview of Smart Power Systems..1

    1.1   The Conventional Power Grid..1
        1.1.1   Overview of a Conventional Power Grid 1
        1.1.2   Problems Associated with Conventional
                   Power Systems 2
                 1.1.2.1 Cascading Failure....................................2
                 1.1.2.2 Environmental Issues 3
    1.2   Future Grid..4
        1.2.1   What Is a Smart Grid?..4
        1.2.2   Smart Grid Characteristics 5
        1.2.3   Main Functionalities of a Smart Grid......................6
        1.2.4   Smart Grid Communication Network 7
        1.2.5   Integration from Supply to Demand in a
                 Smart Grid 7

Chapter 2   Distributed Generation Technology 9

    2.1   Distributed Generation..9
        2.1.1   Introduction 9
        2.1.2   Advantages of Distributed Generation 9
    2.2   Renewable Energy Systems 10
    2.3   Renewable Generation Technologies 11
        2.3.1   Solar Energy 11
                 2.3.1.1 Available Topologies...............................12
                 2.3.1.2 Science behind Solar Energy 12
                 2.3.1.3 PV Efficiency 13
                 2.3.1.4 Solar PV System to Grid..........................13
                 2.3.1.5 Mathematical Model of a Solar
                         PV Cell...14
                 2.3.1.6 From Cells to Modules to Arrays.............18
                 2.3.1.7 Effect of Irradiance.................................18
                 2.3.1.8 Effect of Temperature on I-V
                         Curves 21
        2.3.2   Wind Energy...25
                 2.3.2.1 Basics of Wind Energy.............................25

                  2.3.2.2 Grid Integration: Synchronizing
                             with the Grid 29
                  2.3.2.3 Synchronization Process of Wind
                             Energy Systems 29
        2.3.3   Energy Storage Systems 30
                2.3.3.1 Electrochemical Battery 31
                2.3.3.2 Flywheel 31

Chapter 3 Overview of Microgrids 33

    3.1   What Is a Microgrid? 33
    3.2   Microgrid Power Architecture 34
        3.2.1   Microgrid Structure and Components 34
        3.2.2   Types of Power Architecture 35
    3.3   Operation of Microgrid 36
        3.3.1   Modes of Operation 36
                3.3.1.1 Grid-Connected Mode 36
                3.3.1.2 Islanded Mode 36
        3.3.2   Demand–Supply Balance 37
        3.3.3   Types of Distributed Generators Based on
                Different Operating Conditions 38
                3.3.3.1 Grid-Forming Units 38
                3.3.3.2 Grid-Feeding Units 38
                3.3.3.3 Grid-Following Units 39
        3.3.4   Types of Electrical Load 39
                3.3.4.1 Resistive Loads 39
                3.3.4.2 Capacitive Loads 39
                3.3.4.3 Inductive Loads 39
                3.3.4.4 Combination Loads 40
    3.4   Types of Microgrid Control Architecture 40
        3.4.1   Centralized Control 40
        3.4.2   Decentralized Control 41
        3.4.3   Distributed Control 41
        3.4.4   Hierarchical Control 42
                3.4.4.1 Droop Control 42
                3.4.4.2 Primary Control 43
                3.4.4.3 Secondary Control 44
                3.4.4.4 Tertiary Control 44
    3.5   Advantages and Disadvantages of Microgrids 45
        3.5.1   Advantages of Microgrids 45
        3.5.2   Disadvantages of Microgrids 45
    3.6   Networked Microgrids 45
    3.7   Example: Microgrid Modeling and Simulation 46

Chapter 4   Novel Approaches to Microgrid Functions 65

    4.1   Reconfigurable Power Electronic Interfaces........................65
        4.1.1   Introduction to Power Electronic Interfaces............65
        4.1.2   DC to DC Converters 65
                4.1.2.1   Buck Converter..67
                4.1.2.2   Boost Converter.......................................71
                4.1.2.3   Buck–Boost Converter 72
        4.1.3   DC to AC Inverters 72
                4.1.3.1   Voltage Source Inverter 72
                4.1.3.2   Current Source Inverter 73
                4.1.3.3   Z Source Inverter....................................74
        4.1.4   Reconfigurable Power and Control
              Architectures of Microgrids 74
                4.1.4.1   Reconfigurable Systems...........................74
                4.1.4.2   Existing Power Architecture-Based
                        Reconfigurable Approaches for
                        Microgrids...74
                4.1.4.3   Existing Control Architecture-Based
                        Reconfigurable Approaches for
                        Microgrids...75
        4.1.5   Modeling of Solar Microgrids with a Z Source
              Inverter..75
                4.1.5.1   Example of Proposed System
                        with a ZSI..76
                4.1.5.2   Modes of Control of a ZSI 77
                4.1.5.3   Advantages of a ZSI.................................78
    4.2   Adaptive Protection for Microgrids 79
        4.2.1   Overview of Power System Protection.....................79
                4.2.1.1   Protection System Components 79
                4.2.1.2   Properties of a Protection System 80
        4.2.2   Present Microgrid Protection Schemes 81
                4.2.2.1   Line Protection 81
                4.2.2.2   Primary and Backup Protection 81
        4.2.3   Adaptive Protection Schemes for Microgrids 81
                4.2.3.1   What Is Adaptive Protection?...................82
                4.2.3.2   Adaptive Protection Algorithms...............82
        4.2.4   Case Study 83
    4.3   Multi-Agent-Based Control 86
        4.3.1   Introduction to Multi-Agent Systems......................86
        4.3.2   Multi-Agent-Based Control for Microgrids 88
                4.3.2.1   Proposed System 88
                4.3.2.2   Agents in the System and Their
                        Functions...88

4.3.3   Simulating the Interaction between Agents
Using JAVA Agent Development
Environment 89
     4.3.3.1 JAVA Agent Development
Environment 89
     4.3.3.2 Agent Formation 90
     4.3.3.3 Sniffing Agent 91

Chapter 5   Cyber Security for Smart Microgrids..95

5.1   Overview of Cyber Attacks...95
    5.1.1   Types of Cyber Attack...95
       5.1.1.1 Malware...95
       5.1.1.2 Phishing...95
       5.1.1.3 Man in the Middle Attack 95
       5.1.1.4 Denial of Service Attack 95
       5.1.1.5 Ransomware...96
    5.1.2   Common Sources of Cyber Threats 96
5.2   Power Routing Concept 96
5.3   Cyber Security-Enabled Power Systems.................................97

Chapter 6   Expert Systems for Microgrids..101

6.1   Optimization of Energy Management Systems for
Microgrids Using Reinforcement Learning 101
    6.1.1   Supervised, Unsupervised, and Reinforcement
Learning...101
    6.1.2   Fundamentals of Reinforcement Learning............101
       6.1.2.1 General Reinforcement Learning
Model 101
       6.1.2.2 Markov Decision Process.......................102
       6.1.2.3 The Goal of the Reinforcement
Learning Agent.......................................103
       6.1.2.4 Policies and Value Functions.................104
       6.1.2.5 Sample-Based Learning 105
       6.1.2.6 On- and Off-Policy Learning
Methods..106
       6.1.2.7 SARSA vs Q-Learning...........................107
       6.1.2.8 Q-Learning Algorithm............................107
       6.1.2.9 Exploration and Exploitation
Strategy 108
       6.1.2.10 Hyperparameter Selection......................109
    6.1.3   Single and Multi-Agent Reinforcement
Learning...111

6.1.4   Problem Formulation in RL 112
        6.1.4.1   Defining the Goal....................................112
        6.1.4.2   Mapping the Problem with RL
                  Elements..112
6.1.5   Reinforcement Learning Approach for
        Microgrids 114
        6.1.5.1   Grid Consumption Minimization 114
        6.1.5.2   Minimization of Demand–Supply
                  Deficit..114
        6.1.5.3   Islanded Operation of Microgrids 116
        6.1.5.4   Economic Dispatch 116
        6.1.5.5   Energy Market.......................................117
6.2 Case Study: Reinforcement Learning Approach for
     Minimizing the Grid Dependency of a Solar
     Microgrid 117
     6.2.1   Proposed System 117
     6.2.2   Single-Agent Reinforcement Learning
             Model 119
     6.2.3   Multi-Agent Reinforcement Learning
             Model 121
     6.2.4   Simulation Model 123
             6.2.4.1   Artificial Neural Network 123
             6.2.4.2   Feature Selection...................................125
     6.2.5   RL Simulation Models in Python 125
     6.2.6   Hardware Implementation......................................130
             6.2.6.1   Microgrid Testbed 130
             6.2.6.2   Agent Implementation...........................130
     6.2.7   Agent Communication 134
     6.2.8   Firebase Database...136

Chapter 7   Conclusion...139

Index..145

Preface

Nowadays, technology enhancement, economic and population growth, dependency on digital solutions, and so on, result in the rapid growth of electricity demand. To address the ever-growing demand for electricity, centralized, large-scale, fossil fuel-based power plants were established. This conventional power infrastructure created life-threatening conditions for living beings due to the excessive burning of fossil fuels that emit greenhouse gases such as CO_2 and SO_2. As the concentration of greenhouse gases in the atmosphere increased, an aspect of global climate change, namely "global warming", came into the picture. After finding the root causes of global warming, people became much more concerned about eliminating those factors through prevention and alternative approaches. Environmental concern has changed the energy infrastructure, as countries have developed their nationally determined contributions to clean, green energy sources. Therefore, distributed generation has become an emerging trend.

Distributed generation encourages the interconnection of renewable energy resources with the power network at or near the point of consumption. However, the high penetration of distributed renewable energy resources such as wind and solar photovoltaic (PV) introduces specific challenges to conventional power networks such as a lack of inertia, weather dependency and power fluctuations. Continuous controlling and monitoring are essential to maintain system stability by ensuring the demand–supply balance and power quality. Rather than relying on the centralized control of the main grid, power experts have designed separate integrating platforms, namely "microgrids", for these distributed generation units, with local controllers being available to mitigate the inherent issues with renewable energy generation. Microgrids can operate in multiple modes to provide an uninterrupted power supply to the local loads, regardless of the availability of the utility grid.

The internet era has transformed this new power infrastructure in a somewhat similar, but more effective, manner to the change induced by environmental concerns. As a result, microgrids have become more advanced, with the integration of communication and information technology to improve their operations. Self-healing, interactive, predictive, and secured microgrids have become the newest trend in the power sector.

Many researchers are interested in modifying existing power systems with the smart approach. Instead of centralized control, an agent-based distributed control structure has been introduced for microgrid operations. In a novel approach, these agents are developed with the learning ability to optimize the microgrid's functions. Learning methods introduce intelligent energy storage units to the system to utilize the generation optimally while contributing to the utility grid as peak shavers. In addition, the seamless multimode operation of the microgrid requires the operation of mode-based power electronic interfaced units and protection, which can be achieved through reconfigurable power system components and adaptive protection methods, respectively. Likewise, the smart concept has the ability to enhance microgrid performance through innovative solutions. The underlying communication infrastructure of a smart microgrid is its backbone and the network that is most vulnerable to threats. It is crucial to protect against cyber attacks such as illegal data acquisition, unauthorized access, server failures, and so on, and to promote the power routing concept for a flexible and dynamic distribution network with intelligent controllers.

This book aims to provide comprehensive coverage of previously discussed novel technologies. These new technologies are based on the research experiences gained by the authors over the past decade. This is a handbook on smart technologies for microgrid operations for undergraduate, graduate and postgraduate students and any other researchers or readers interested in the theoretical background and related literature, the design and implementation, the analysis of results, and the outcomes for future power networks.

Initially, the book enlightens the reader on the smart power system concept by covering its functionalities in the areas of transmission, distribution, and utilization. In Chapter 2, the authors present essential background knowledge on distributed generation technologies and hybrid power systems. Following that, Chapter 3 provides a complete analysis of microgrid design architecture and implementation, based on the authors' research experience. Case studies related to microgrid modeling and simulation are provided at the end of the chapter. These present useful guidelines for new researchers in this sector who are designing and developing innovative models.

Chapter 4 emphasizes smart approaches to microgrids, based on protection, control and communication architecture, and power electronic interfaces for distributed generation integration. Sufficient weight is given in this book to microgrid protection and fault studies, which are the foundation for ensuring the reliable operation of the system. As a novel approach for future power systems, the concept of an adaptive protection scheme is introduced with relevant research-based approaches. Furthermore, an overview of power electronic interfaces is provided. In addition to the conventional converter and inverter architecture, a reconfigurable interface is introduced. This chapter also presents a distributed control architecture based on multiple agents. These novel smart approaches to regular microgrid functions are supported by relevant case studies and simulation models to provide readers with a more practical and reliable background.

The authors have enriched the comprehensive coverage of smart technologies by adding a separate chapter on a practical challenge for smart power infrastructure, which is the enabling of cyber security. Chapter 5 highlights the importance of, and approaches to, mitigating cyber threats. Chapter 6 introduces expert systems on smart microgrids where system optimization is achieved through intelligent approaches. The inclusion of machine learning based approaches for microgrids, with relevant case studies and simulation models, ensures that the contents of the book are of high quality and up to date. The authors' conclusion paves the way for the exploration of new research gaps and trending concepts related to smart approaches to microgrids, to encourage budding researchers and interested readers.

Acknowledgements

The authors convey their heartfelt gratitude to the members and colleagues of the Smart Grid Research Group of the Department of Electrical Engineering at the University of Moratuwa for their support in the successful completion of this book.

Authors

KTM Udayanga Hemapala earned a BSc (Eng) at the University of Moratuwa, Sri Lanka, in 2004 and a PhD at the University of Genova, Italy, in 2009. He is Professor in the Department of Electrical Engineering, University of Moratuwa. His research interests include industrial robotics, distributed generation, power system control and smart grids.

MK Perera earned a BSc (Eng) at the University of Moratuwa, Sri Lanka, in 2020. She is a postgraduate student and a research assistant in the Department of Electrical Engineering, University of Moratuwa. Her research interests include renewable generation, smart grids and machine learning.

Introduction

In order to address the rapid increase in electricity demand, power producers were highly dependent on fossil fuel-based, large-scale power plants. The burning of fossil fuels releases greenhouse gases. The heavy reliance on fossil fuels resulted in an increase in the greenhouse gas concentration in the atmosphere and gave rise to global warming phenomena. As a result, power experts became much more concerned about mitigating the environmental impact of electricity generation. This has become the motivation for distributed renewable generation. Renewable generation penetration requires a new power infrastructure to overcome its inherent issues such as dependency on weather, lack of inertia, and energy storage requirements.

This book highlights how microgrids have been identified as integrated platforms for distributed generation units, energy storage and local loads. Furthermore, it emphasizes how the performance of such a system can be improved through innovative approaches. The book discusses, using research outcomes, the development of agent-based control systems to break down complex control tasks into smaller tasks, the introduction of machine learning-based optimization techniques for energy management functions, renewable generation forecasting, energy storage management for peak shaving, and the optimum utilization of renewable generation.

The authors aim to present comprehensive coverage of all aspects of microgrid modeling, so microgrid protection and reconfiguration are also addressed in this book. Unlike conventional power systems, microgrids support multimode operations in which the continuous monitoring of system states, and protective responses to adjust relay settings accordingly, are essential. The requirements of adaptive protection schemes are highlighted here. In addition, the book presents sufficient background knowledge on the power electronic interfaces that are essential for distributed generation units. As a novel concept, the reconfiguration of power electronic interface units to respond automatically and switch to the optimum operating mode is presented as a research-based outcome.

The core of this book is advanced technologies for smart microgrid systems, which is the integration of all the previously discussed functions through an underlying communication network. The transformation of the modern power infrastructure in the internet era has resulted in a new set of challenges, namely, cyber attacks including access, reconnaissance, and password attacks. An insight into mitigating the effects of these attacks is also available as a foundation to the power routing concept.

1 Overview of Smart Power Systems

1.1 THE CONVENTIONAL POWER GRID

1.1.1 OVERVIEW OF A CONVENTIONAL POWER GRID

A conventional power grid consists of four main sections, as shown in Figure 1.1, namely generation, transmission, distribution, and utilization. Conventional power systems are based on large, centralized power generating stations. In most of these power stations, the electricity is generated by burning a fossil fuel and producing steam, which is then used to drive a steam turbine that in turn drives an electrical generator. In addition, conventional generation relies on nuclear power plants or major hydro plants where the capacity of the plant is in the MW range. The generated energy is delivered to the end user through complex transmission and distribution systems. Here, step-up transformers are used to increase the voltage to transmit the electricity to consumer centers, then the voltage is reduced for distribution using step-down transformers. Finally, the electricity comes to the consumer at a specified voltage and with the specified frequency at minimum cost. This process appears seamless to the end user. For example, when you flip a switch the electrical equipment is turned on. However, the existing power network is often described as the most complex machine ever built. It is a real-time energy delivery system. The main goal of the power system is to cater for loads with electricity at the specified voltage and frequency at minimum cost consistent with operating constraints, safety, and so on.

The main characteristics of a conventional power network can be described as follows:

- High power plant capacity
- Lengthy transmission network
- Uni-directional power flow
- Low load controllability

There are several challenges faced by power system operators in achieving the required performance. The main reason for these challenges is the real-time

DOI: 10.1201/9781003216292-1

1

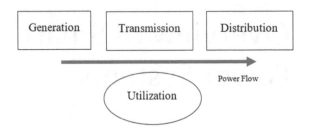

FIGURE 1.1   General structure of a conventional power grid

operation of the electric power system. Real-time means that power is generated, transported, and supplied the moment you turn on some electrical equipment. In addition to that, a power system is subjected to disturbances, such as short circuit faults and lightning strikes, so a properly designed and coordinated power system protection scheme is required. In achieving the main goals of an existing power network, there are engineering tradeoffs between reliability and cost.

1.1.2 Problems Associated with Conventional Power Systems

There are certain problems associated with conventional power systems, as they are based on centralized generating stations. Bulk generation, heavy reliance on fossil fuels and cascading failures can be highlighted as examples.

1.1.2.1 Cascading Failure

Existing power networks are subject to the risk of cascading failures. What is a cascading failure? It is a failure in a system of interconnected parts in which the failure of one part triggers the failures of successive parts. This may result in power interruptions. For example, in an industry, if there is a grid power interruption there is no, or reduced, production, resulting in a loss of revenue. Several examples of cascading failures can be given as follows.

1.1.2.1.1 Fukushima Nuclear Power Plant Accident

This is good example of a cascading failure, where the sequence of events in Figure 1.2 happened in the plant after a tsunami hit the area.

1.1.2.1.2 Blackout in Sri Lanka (2020)

After the blackout in 2020, which commenced at around 12.30pm on 17 August and lasted for between 7 and 8 hours, the Minister of Power of Sri Lanka appointed a committee (comprising two administrative officers, an additional secretary in the Ministry of Power who served as the chairman, a retired professor of mechanical engineering, an engineer who was the chairman of a corporation, two senior lecturers in electrical engineering, one senior official from the Ceylon Electricity Board (CEB), and one senior official from the Ministry of Power who

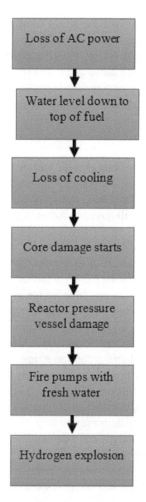

FIGURE 1.2   Sequence of events in Fukushima nuclear power plant accident

was responsible for renewable energy development) to investigate the matter and submit a report within a week. As described in the committee report, one of the important grid substations tripped, so the system frequency increased beyond the current setting for the rate of frequency, tripping the relay of the main coal power plant, Lak Vijaya Power Station (LVPS). As a result, the generator-transformer circuit breakers of all three units of the LVPS made the power plant unavailable to the grid, resulting in a cascading failure.

1.1.2.2   Environmental Issues
Heavy reliance on fossil fuels causes emissions of carbon dioxide, which result in life-threatening conditions as they enhance the greenhouse effect. It is estimated

that the annual CO_2 emissions from the world's energy-related industries total 20 billion tonnes. Greenhouse gas molecules trap heat waves that reflect from the Earth's surface and reradiate them towards the Earth. This disturbs the natural removal of heat from the Earth's atmosphere and results in an overall temperature rise. To find solutions for the greenhouse effect, the world is now more concerned about reducing fossil fuel and promoting renewable generation.

1.2   FUTURE GRID

The problems stated above associated with conventional power systems are the main drivers for changing the existing energy infrastructure. Therefore, the existing conventional grid needs to be updated, and the future grid will be more complex than today's. The future grid will have the following characteristics:

- Decentralized power system with more distributed generation: The distributed nature of a smart grid ensures the integration of decentralized power generation such as solar panels, wind turbines, and biomass generators. The future power system needs to integrate green technologies, as this has now become a compulsory requirement due to global warming and other environmental issues.
- Consumers become producers (Prosumers): A prosumer is an individual who both consumes and produces at the same time.
- Multi-directional power flow: Once consumers become producers, the power flow will be in both directions. The system operators should then have the capability to control the variable distributed generation based on renewable energy. Otherwise, the system may be unstable.
- Flexible loads from participating in demand response programs.
- Adaptive and islanding capabilities.
- Self-healing capabilities: Self-healing refers to the ability to identify power grid problems in real time and safely respond by self-correcting.

When a grid is modernized with one or more of the characteristics listed above, it is converted to a modern grid and referred to as a "smart grid".

1.2.1   What Is a Smart Grid?

A smart grid is an electrical system integrated with communications and information technology for enhanced grid operations, customer services, and environmental benefits.

For example, a new digital meter on your breaker panel will be a smart device and the initial stage of the realization of a smart grid. Adding a wireless network that reads that meter remotely or a data management system that processes the information will add more benefits. Adding some solar panels on the roof will then be beneficial in terms of distributed generation. The customer may then need some sort of a load controller for their heating, ventilation, and air conditioning system. All of these come as part of a smart grid.

1.2.2 SMART GRID CHARACTERISTICS

The characteristics of a smart grid are presented in Figure 1.3.

The distributed nature of a smart grid ensures the integration of decentralized power generation such as solar panels, wind turbines, and biomass generators. Also, smart grids combine intelligence and control, for the optimum utilization of power production and energy storage resources. The predictive nature of a smart grid refers to the measurement and analysis of anomalies that affect power quality, to prevent emergencies from occurring. In addition, the inherent features of a smart grid can be listed as follows:

- Use of automation
- Distributed energy delivery network
- Possibility of two-way flow of electricity
- Possibility of two-way flow of information
- Capability to monitor and respond to changes in the power system
- Ability to identify and resolve faults on the electricity grid
- Monitoring of power quality and managing voltage
- Identification of devices or subsystems that require maintenance
- Help for consumers to optimize their individual electricity consumption (minimize their bills)
- Enabling the use of smart appliances that can be programmed to run on off-peak power

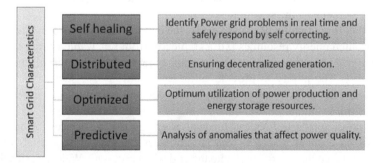

FIGURE 1.3 Smart grid characteristics

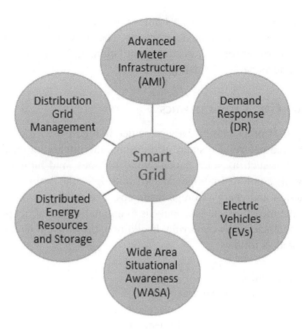

FIGURE 1.4   Functionalities of a smart grid

1.2.3   MAIN FUNCTIONALITIES OF A SMART GRID

The main functionalities of a smart grid are shown in Figure 1.4.

Advanced metering infrastructure (AMI) allows utilities to collect, measure, and analyze energy consumption data for various purposes such as grid management, outage notification, and billing, via two-way communications. One of the most common steps taken by utilities when creating a smarter power grid is the increasing implementation of demand response (DR). Demand response is the reduction in the use of electric energy by customers in response to an electricity price increase or heavy burdens on the system.

Another functionality of a smart grid is the better and more uniform integration of distributed energy resources (DERs), mostly from renewable energy sources, into the grid. Electric vehicles (EVs) contribute to the reduction of emissions and energy. The ability to provide sufficient electricity for such vehicles will depend on effectively managing supply and demand, which is a core benefit of a smart grid.

In a novel approach, EVs can function as energy storage devices for balancing demands on a smart grid. EVs can absorb excess supply during periods of low demand and feed that energy back into the grid when necessary. Wide area situational awareness (WASA) refers to the implementation of a set of technologies to improve the monitoring of a power system across large geographical areas. This provides grid operators with a broad and dynamic picture of the functioning of the grid.

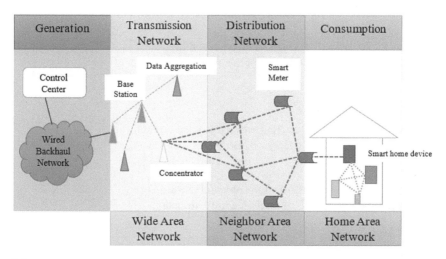

FIGURE 1.5 Smart grid communication network

1.2.4 SMART GRID COMMUNICATION NETWORK

The key to achieving these potential benefits of a smart grid is a properly designed smart grid communication network (SGCN) that can support all the identified smart grid functionalities. The general structure of an SGCN is given in Figure 1.5. In a smart grid, in parallel to the power infrastructure there is an underlying communication network that is the backbone of the smart grid.

There are three communication network architecture layers based on a wide area network (WAN), a neighbor area network (NAN) and a home area network (HAN).

A home area network supports home/building/industrial automation-specific applications in sending or receiving data sensed from electrical appliances to or from the built-in controllers of the appliances. The main requirements are low power consumption, low cost, ease and secure links for communication.

A neighbor area network supports smart metering and distribution-specific applications in sending or receiving the data transmitted from customer/field devices to or from the substation/concentrator. The main requirements are a high data rate and a wide geographical coverage.

A wide area network supports wide-area control, protection and monitoring to transmit a huge amount of data at a much higher data rate and over the longest distance.

1.2.5 INTEGRATION FROM SUPPLY TO DEMAND IN A SMART GRID

The future of smart grids is as a new value chain augmented and interconnected by digital technologies and leading to a digital transformation era in which both power and information flow in multiple directions. In the last few years, many countries

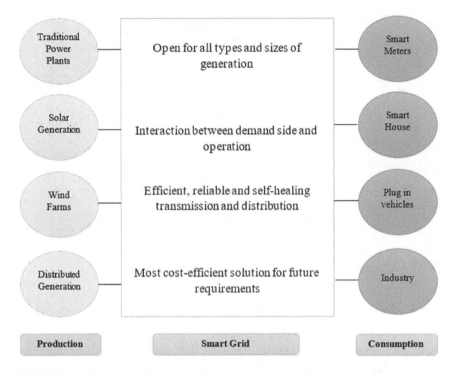

FIGURE 1.6 Summary of integration from supply to demand in a smart grid

in the world have shown a huge interest in smart grid technology. They are facing many challenges in the process of deployment of this technology at ground level. It has been identified that research and development are key to achieving the reality of a smart grid within a country. A summary of the integration from supply to demand in a smart grid is given in Figure 1.6.

2 Distributed Generation Technology

2.1 DISTRIBUTED GENERATION

2.1.1 INTRODUCTION

Distributed generation (DG) relates to various generating technologies that generate electricity at or near a utility. It is also referred to as cogeneration or small power production. Even home-installed solar panels, emergency generators, and biomass-based electricity generators can be considered as distributed generation units. As shown in Figure 2.1, through proper regulation and conversion, distributed generation resources such as solar PV systems, wind turbines, diesel generators and battery backups can be integrated to cater for nearby loads.

Examples:

- Micro turbines
- Fuel cells
- PV systems
- Wind turbines
- Combined heat power (CHP) systems

To accommodate variable generation, operators must ensure that there is sufficient flexibility in the rest of the system to keep the system in balance. System flexibility can come from a number of sources, including spinning reserves, existing generator ramping capability, power flows between balancing areas, demand response, and energy storage.

2.1.2 ADVANTAGES OF DISTRIBUTED GENERATION

Most importantly, distributed generation saves energy costs by reducing the contribution of centralized power plants. It also reduces transmission and distribution losses by producing energy at or near the point of consumption. Distributed generation encourages the use of non-conventional renewable energy sources such as wind, solar photovoltaic, mini hydro, and biomass. In addition,

DOI: 10.1201/9781003216292-2

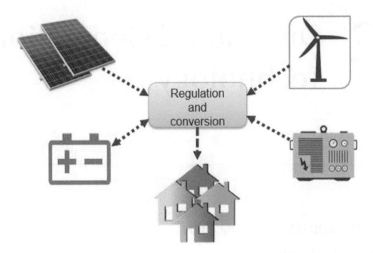

FIGURE 2.1 Distributed generation concept

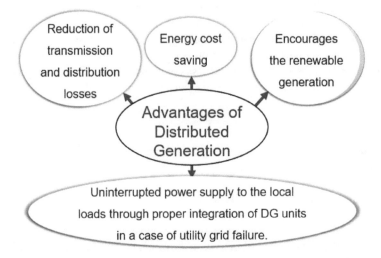

FIGURE 2.2 Advantages of distributed generation

in the event of a failure of the utility grid, properly integrated distributed energy resources can maintain an uninterrupted power supply to the consumers within a defined electrical boundary. A summary of the advantages of distributed generation is given in Figure 2.2.

2.2   RENEWABLE ENERGY SYSTEMS

Energy can be generated mainly from the conversion of the chemical energy in a fuel or by utilizing the flowing energy of wind, water, or steam. The energy

Primary Global Energy Consumption 2020

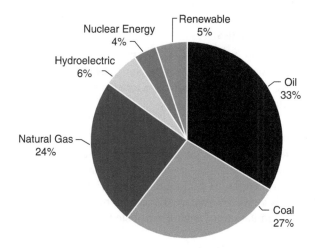

FIGURE 2.3   World energy usage. (Data source: BP Statistical Review 2020.)

generated can be used to produce steam to operate a generator, or can directly operate a generator to produce electrical energy.

Energy sources can basically be categorized as non-renewable and renewable energy sources. International and local organizations who are mostly concerned about environmental sustainability promote the use of renewable energy, but most countries and power utilities are still using fossil fuel based centrally controlled power plants, because of the technical advantages. The following Figure 2.3 shows the world energy usage in the different categories.

This clearly shows that the use of non-renewable energy sources such as fossil fuels still remains higher. Therefore, in order to achieve the renewable energy targets, there is a need to study the barriers to the implementation of renewable energy systems. Therefore, the following sections will discuss the technical side of the main renewable energy technologies, including solar PV and wind energy systems.

2.3   RENEWABLE GENERATION TECHNOLOGIES

2.3.1   SOLAR ENERGY

Solar energy is the main source for most renewable energy systems. It is free, inexhaustible and completely pollution-free, but there are also a few drawbacks. For some systems the energy density per unit area is very low, the energy is available for only part of the day, and cloudy and hazy atmospheric conditions greatly reduce the energy received. To harness solar energy for electricity generation, there are challenging technological problems. The most important of these is the collection and concentration of solar energy and its conversion to the electrical form through efficient and comparatively economical means.

2.3.1.1   Available Topologies

2.3.1.1.1   Concentrated Solar-Thermal Power

These systems concentrate the radiation of the Sun to heat a liquid substance which is then used to drive a heat engine and drive an electric generator. This indirect method generates alternating current (AC).

2.3.1.1.2   Photovoltaic Systems

These systems differ from solar thermal systems in that they do not use the Sun's heat to generate power. Instead, they use sunlight and the "photovoltaic effect" to generate direct electric current (DC) in a direct electricity production process by solar cells. The DC is then converted to AC, usually by using inverters.

There are several limitations of solar PV, as listed below:

- Low efficiency (12% to 18%)
- High initial investment
- Plant availability (plant factor, which is the ratio of the actual energy output of the power plant over a period of time to its potential output if it had operated at full nameplate capacity the entire time, of about 15% to 25%)
- High space requirement
- High dependencies on geographical limitations, weather and atmospheric conditions

2.3.1.2   Science behind Solar Energy

It is important to know the properties of the Sun. The total power output from the Sun is 3.8×10^{26} W. The power density can be calculated by dividing the total power by the surface area of the Earth, as in equation (2.1). The distance between the Sun and the Earth is 1.49×10^{11} m.

$$\frac{P}{A} = \frac{3.8 \times 10^{26}\,W}{4\pi \times \left(1.49 \times 10^{11}\,m\right)^2} = 1367W / m^2 \tag{2.1}$$

The Earth's radius is 6.37×10^6 m. The total solar power incident on the Earth is given by the power density times the apparent area of the Earth, as in equation (2.2),

$$P_{total} = \left(1367W / m^2\right) \times \pi \times \left(6.37 \times 10^6\,m\right)^2 = 1.73 \times 10^{17}\,W \tag{2.2}$$

The average insolation at the Earth's surface is given by equation (2.3),

$$\frac{P}{A} = \frac{0.5 \times 1.73 \times 10^{17}\,W}{4\pi \times \left(6.371 \times 10^6\,m\right)^2} = 168W/m^2 \tag{2.3}$$

This is a simple calculation to show how the insolation can be calculated, but the insolation at any given location on the Earth at any given time depends on the time of day, the day of the year and the location.

2.3.1.3 PV Efficiency

Table 2.1 shows the maximum theoretical efficiency of PV cells made of various semiconductor materials. Silicon-based cells have a typical efficiency of 15–17%, with higher-quality cells reaching up to 23%.

Let's consider the following examples in relation to PV efficiency.

Calculate the cost per kWh, averaged over an operational period of 10 years, for a photovoltaic system with an installation cost of $3.00 per W, if the system operates with a capacity factor of 8% (this is the net result of the photovoltaic efficiency and the fraction of time that is daylight). (Do not include the cost recovery factor.)

Using the average (cloud-free) insolation on Earth as calculated previously, calculate the area of a horizontal photovoltaic array with an efficiency of 20% that would be needed to satisfy the residential electricity needs of a city of 250,000 people with an average of 2.6 people per household.

2.3.1.4 Solar PV System to Grid

PV systems are highly modular, with power output varying from a few watts to tens of megawatts. As shown in Figure 2.4, an inverter converts the DC output from the PV system to AC and connects it to the grid via an isolating transformer. PV systems can be connected to the grid as distributed generators, allowing for bi-directional power flow in the network.

TABLE 2.1
Maximum Theoretical Efficiency of PV Cells Based on Different Semiconductor Materials

Material	Maximum Theoretical Efficiency (%)
CdS	18
AlSb	27
CdTe	27
GaAs	26
InP	26
Si	25
Ge	13

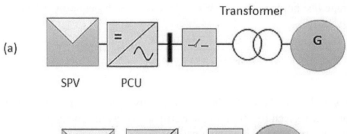

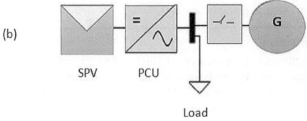

FIGURE 2.4 Solar PV system to grid

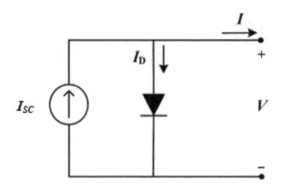

FIGURE 2.5 Simplified model of PV cell

2.3.1.5 Mathematical Model of a Solar PV Cell

The p-n junction of a semiconductor material that creates electricity when exposed to light is the most fundamental component of a photovoltaic cell. As shown in Figure 2.5, a simple equivalent circuit model for a PV cell consists of a real diode in parallel with an ideal current source. For both the actual PV and its equivalent circuit, there are two requirements of significant importance:

1. The current that flows when the terminals are shorted together (the short circuit current, I_{sc})

2. The voltage across the terminals when the leads are left open (the open circuit voltage, V_{oc})

We can derive equation (2.4) for the above equivalent circuit:

$$I_{SC} = I_d + I \tag{2.4}$$

The p-n junction characteristics can be used to examine the PV cell voltage and the current characteristics, as shown in the simplified circuit model above.

The voltage–current characteristic curve for a p-n junction diode is defined by the well-known Shockley Diode equation, as shown in equation (2.5).

$$I_d = I_0 \left[e^{\frac{qV_d}{AkT}} - 1 \right] = I_0 \left[e^{\frac{V_d}{AV_T}} - 1 \right] \tag{2.5}$$

I_d = *current through the diode* (A)

V_d = *diode voltage with anode positive with respect to cathode* (V)

I_0 = *leakage* (*reverse saturation*) *current* (A)

A = *empirical constant known as the emission coefficient*

q = *electron charge* 1.6022×10^{-19} *Coulomb*

T = *Absolute temperature in Kelvin*

k = *Boltzman constant* $\left(1.381 \times 10^{-23} \dfrac{J}{K} \right)$

$$\textit{Thermal voltage } V_T = \frac{kT}{q} = \frac{1.381 \times 10^{-28} \times T}{1.602 \times 10^{-19}} V \tag{2.6}$$

For a junction temperature of $25°C$, $I_d = I_0 \left[e^{\frac{qV_d}{AkT}} - 1 \right] = I_0 \left[e^{38.9V_d} - 1 \right]$ $\quad$ (2.7)

By substituting equation (2.7) in equation (2.4), we get equation (2.8).

$$I = I_{SC} - I_0 \left[e^{\frac{qV_d}{AkT}} - 1 \right] \tag{2.8}$$

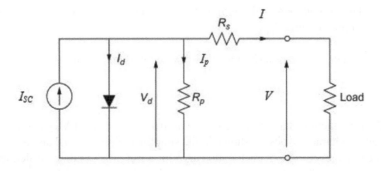

FIGURE 2.6 More accurate model of a PV cell with parallel and series resistances

When the leads from the PV cell are left open, $I = 0$ and we can derive equation (2.9) for the open circuit voltage V_{OC}.

$$V_{OC} = \frac{kT}{q} \ln \left[\frac{I_{SC}}{I_0} + 1 \right] \tag{2.9}$$

Instead of using the most basic solar PV model, we consider a more accurate model that includes a series (R_S) and parallel (R_P) resistance in the PV system model, resulting in the PV system equivalent circuit model shown in Figure 2.6. The contact resistance associated with the link between the cell and its wire leads, as well as the resistance of the semiconductor itself, are referred to as series resistance. The leakage currents surrounding the edge of the PV cell are represented by the parallel resistance. The PV cell current I and voltage V can be calculated using equations (2.10) and (2.11) respectively.

$$I = I_{SC} - I_d - I_P$$

$$I = I_{SC} - I_0 \left[\exp \left(\frac{q(IR_S + V)}{AkT} \right) - 1 \right] - \left[\frac{V_d}{R_P} \right] \tag{2.10}$$

$$V = V_d - IR_S \tag{2.11}$$

There is no explicit solution for either the voltage V or the current I in equation (2.10) because it is a complex equation. A spreadsheet solution, on the other hand, is relatively simple. The method is based on incrementing the values of the diode voltage V_d in a spreadsheet. The associated values of the current I and voltage V for each value of V_d are easily found.

Example:

A PV module is made up of 36 identical cells, all wired in series. With 1-Sun insolation (1 kW/m²), each cell has a short circuit current $I_{SC} = 3.4$ A and at 25 °C its reverse saturation current is $I_o = 6 \times 10^{-10}$ A. The parallel resistance Rp = 6.6 Ω and series resistance Rs = 0.005 Ω.

a. Find the voltage, current, and power delivered by the module when the junction voltage of each cell is 0.50 V.
b. Set up a spreadsheet for I and V.
c. Discuss your result in terms of the maximum power output and power rating of the module.

Solution:

a. Using the equation

$$I = I_{SC} - I_0 \left[e^{38.9 V_d} - 1 \right] - \left[\frac{V_d}{R_P} \right]$$

$$I_{cell} = 3.4 - 6 \times 10^{-10} \left(e^{38.9 \times 0.5} - 1 \right) - \frac{0.5}{6.6} = 3.156A$$

$$V = V_d - IR_S$$

$$V_{cell} = 0.5 - 3.156 \times 0.005 = 0.4842V$$

b.

	Calibri	11	A A				Wrap Text	
Paste	B I U ▾ ⊞ ▾ ◇ ▾ A ▾						Merge & Center ▾	

C12   fx   =3.4-0.0000000006*(EXP(38.9*B12)-1)-B12/6.6

	A	B	C	D	E	F
1						
2						
3				Series- Connection		
4						
5						
6		Vd	Module Current_I (A)	Module Voltage_n*V(V)	Power (W)	
7		0.45	3.308	15.605	51.617	
8		0.46	3.295	15.967	52.609	
9		0.47	3.277	16.330	53.506	
10		0.48	3.250	16.695	54.261	
11		0.49	3.212	17.062	54.802	
12		0.5	3.156	17.432	55.020	
13		0.51	3.075	17.807	54.754	
14		0.52	2.956	18.188	53.756	
15		0.53	2.780	18.580	51.655	
16		0.54	2.522	18.986	47.885	
17		0.55	2.142	19.414	41.587	
18						
19						
20						

c. The maximum power point for this module is at $I = 3.156A$ and $V = 17.432V$ and $P = 55.02W$. This module can be described as a $55W$ module.

2.3.1.6   From Cells to Modules to Arrays

An individual cell produces only about 0.5 V, so it is a rare application for which just a single cell is of any use. Instead, the basic building block for PV applications is a module consisting of a number of prewired cells in series, all encased in tough, weather-resistant packages.

Multiple modules can, in turn, be wired in series to increase voltage and in parallel to increase current, the product of which is power. An important element in PV system design is deciding how many modules should be connected in series and how many in parallel to deliver whatever energy is needed. Such a combination of modules is referred to as an array.

Example: Let's observe and compare the changes of output voltage, current, and power for different solar PV configurations using a MATLAB solar PV module as in Figures 2.7 and 2.8.

Case 1: 1 string (with 10 modules connected in series)

Case 2: 2 strings (with 10 modules connected in series)

A combination of PV modules forms a PV array. If the modules are wired in series the total voltage of the array increases in comparison to the voltage of a single module, while the current through all the modules remains the same. If the modules are connected in parallel the total current output of the array increases in comparison to a single module. However, in general arrays are combinations of modules connected in series and modules connected in parallel to improve the power output.

In the first array with a single string (with 10 modules connected in series), the I-V curves of a single module are added along the voltage axis. This increases the open circuit voltage of the final system.

In the second array, two such strings are connected in parallel. The two I-V curves of the first single string system are added along the current axis; this increases the short circuit current of the system and also doubles the power output. However, the voltage across the strings remains the same in both cases.

The results discussed here can be tabulated as in Table 2.2.

Maximum Power Point:

The maximum power is generated by the solar array at a point of the current–voltage characteristic curve where the product VI is at its maximum, as in Figure 2.9.

2.3.1.7   Effect of Irradiance

The higher the irradiance, the higher the generated current in the cell, and the higher the short circuit current. By contrast, the effect on the open circuit voltage is relatively small, as the dependence on the light generated current is logarithmic.

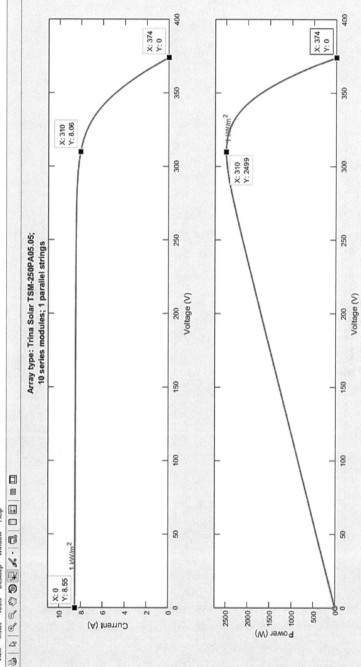

FIGURE 2.7 1 String (with 10 modules connected in series)

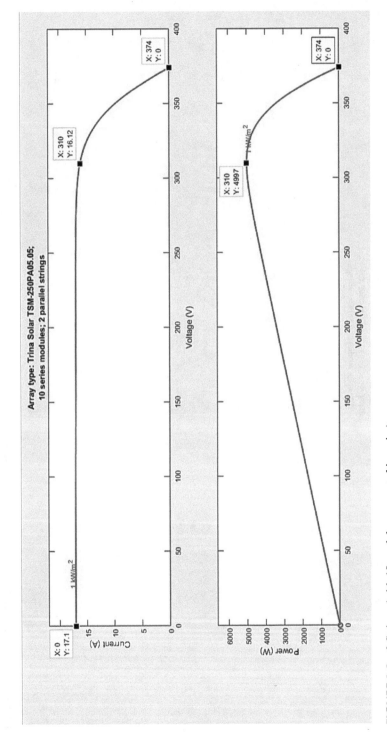

FIGURE 2.8 2 Strings (with 10 modules connected in series)

TABLE 2.2
Comparison between Results Obtained from Two Different Solar Array Configurations

	1 String	2 Strings
Maximum Power (P_max)(W)	2499	4997
Maximum Power Point Voltage (V_mpp) (V)	310	310
Maximum Power Point Current (I_mpp) (A)	8.06	16.12
Open Circuit Voltage (V_oc) (V)	374	374
Short Circuit Current (I_sc) (A)	8.55	7.10

Example: Let's consider the MATLAB Simulink model given in Figure 2.10. Let's observe and compare the changes of output voltage, current, and power of the solar PV module for six irradiance levels from 1000 W/m^2 to 3500 W/m^2.

2.3.1.7.1 I-V and P-V Variations of the Module with the Irradiance Level at a Temperature of 20 °C

Here six irradiance levels are selected: 1000 W/m^2, 1500 W/m^2, 2000 W/m^2, 2500 W/m^2, 3000 W/m^2 and 3500 W/m^2.

Solar irradiance is the power per unit area received from the Sun in the form of electromagnetic radiation in the wavelength range of the measuring instrument. As the irradiance drops, the short circuit current of the module drops. As we can observe from Figure 2.11, when the irradiance varies from 3 kW/m^2 to 1.5 kW/m^2, the short circuit current of the module drops by half. Thus we can say there is a directly proportional relationship between the irradiance level and the short circuit current. This also changes the open circuit voltage level. As the irradiance level drops, the open circuit voltage drops. This may also result in a decrease in the maximum power output of the module, which can be seen in Figure 2.12 with the variation of the maximum power point.

2.3.1.8 Effect of Temperature on I-V Curves

As the cell temperature increases, the open circuit voltage decreases substantially, while the short circuit current increases only slightly. Therefore, photovoltaics perform better on cold, clear days than on hot days.

Consider again the MATLAB Simulink model given in Figure 2.10. Let's observe and compare the changes in output voltage, current, and power of the solar PV module for six temperature levels from 20 °C to 45 °C.

2.3.1.8.1 I-V and P-V Variations of the Module with the Temperature Level at an Irradiance of 1000 W/m^2

Here six temperature levels are selected: 20 °C, 25 °C, 30 °C, 35 °C, 40 °C, and 45 °C. As the temperature of the cell increases, the open circuit voltage of the module decreases. According to the I-V plot obtained in Figure 2.13, as the temperature

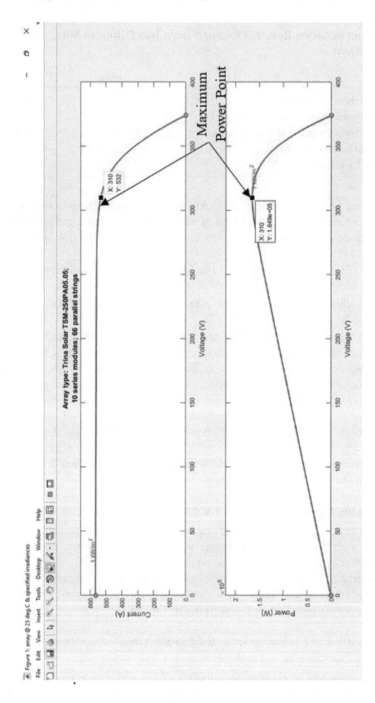

FIGURE 2.9 Maximum power point of a solar PV array

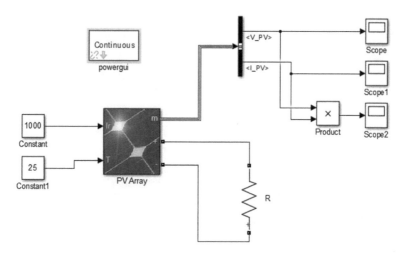

FIGURE 2.10 MATLAB simulink model for a solar PV system

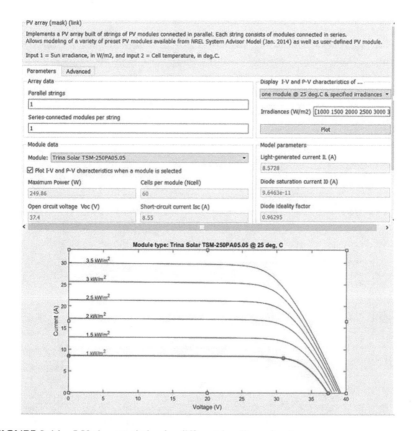

FIGURE 2.11 I-V characteristics for different irradiance levels

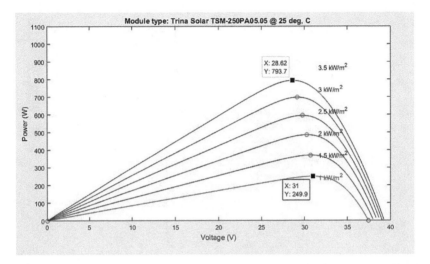

FIGURE 2.12   P-V characteristics for different irradiance levels

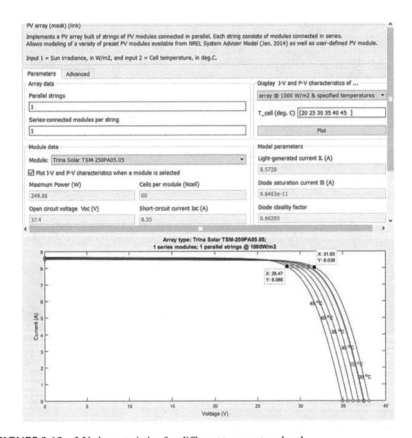

FIGURE 2.13   I-V characteristics for different temperature levels

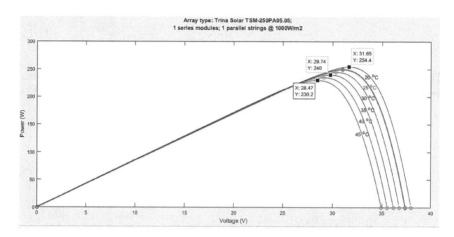

FIGURE 2.14 P-V characteristics for different temperature levels

increases, the short circuit current increases, but this is a slight increase, as shown in the data labels in the I-V plot. As a result, the maximum power output of the module gradually decreases, as indicated by the data labels in the P-V plot in Figure 2.14. Therefore, we can conclude that solar modules perform better in cold weather than on a hot day.

The operating curve of the solar PV module under standard test conditions is indicated by the 1 kW/m² solar irradiance level. Standard test conditions are available to compare the performances of different modules.

(Standard Test Conditions: Solar Irradiance = 1000 W/m², Cell Temperature = 25 °C, Air Mass Ratio = 1.5.)

For crystalline silicon cells, the open circuit voltage drops by about 0.37% for each degree Celsius increase in temperature, and I_{sc} increases by approximately 0.05%. The net result when cells heat up is that the maximum power point (MPP) slides slightly upward and toward the left, with a decrease in the maximum power available of about 0.5% for each degree Celsius. Given this change in performance as the cell temperature changes, it should be quite apparent that temperature needs to be included in any estimate of module performance. Cells vary in temperature not only because the ambient temperature changes, but also because the insolation on the cells changes.

2.3.2 WIND ENERGY

2.3.2.1 Basics of Wind Energy

2.3.2.1.1 Theoretical Wind Power

The theoretical wind power is given by equation (2.12), where ρ is the air density (kg/m³) (at 15 °C and 1 atm, ρ = 1.225 kg/m³); A is the cross-sectional area through which the wind passes (m²); and v_w is the wind speed normal to A (m/s).

$$P_W = \frac{1}{2} A\rho\left(v_W\right)^3 \tag{2.12}$$

The turbine blade captures only a part of the available wind power, and the real mechanical power extracted by a wind turbine is calculated using equation (2.13):

$$P_M = P_W C_P = \frac{1}{2} A C_P \rho\left(v_W\right)^3 W \tag{2.13}$$

The coefficient of performance, sometimes known as the power coefficient, is C_P. A modern turbine's power coefficient is normally between 0.2 and 0.5.

2.3.2.1.2 Tip Speed Ratio

The tip speed ratio is the speed at which the outer tip of the blade is moving, divided by the wind speed. The TSR is given by equation (2.14):

$$TSR = \lambda = \left(rotor\ tip\ speed\right)/\left(wind\ speed\right) = \omega_m R/v_W \tag{2.14}$$

where ω_m is the rotational mechanical speed of the turbine rotor (generator speed) in mechanical rad/s, and R is the radius and $D(=2R)$ is the diameter of the turbine. Thus, the wind speed is given by equation (2.15):

$$v_w = \frac{\omega_m R}{\lambda} \tag{2.15}$$

Substituting equation (2.15) into equation (2.13):

$$P_M = \frac{1}{2}\rho A C_P \left(\frac{\omega_m R}{\lambda}\right)^3 \tag{2.16}$$

2.3.2.1.3 Power Characteristics of a Wind Turbine

The power plot in Figure 2.15 defines the power characteristics of a wind turbine (mechanical power vs wind speed). The power curve is a manufacturer-guaranteed certificate of performance for a wind turbine.

Cut-in wind speed: This is the rate at which the wind turbine begins to operate and produce electricity.

Rated wind speed: The wind turbine produces nominal power at the rated wind speed, which is also the generator's rated output power.

Cut-out wind speed: This is the maximum wind speed at which the turbine can operate before shutting down. The turbine must be stopped if the wind speed exceeds the cut-out speed, to prevent damage from excessive wind.

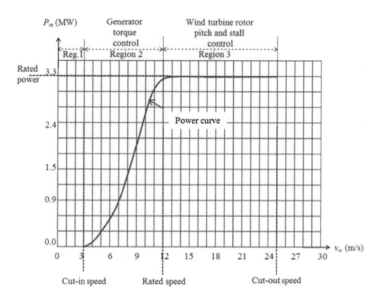

FIGURE 2.15 The power characteristic curve of a wind turbine

2.3.2.1.4 Wind Energy System Configurations

Figure 2.16 shows the different configurations of wind energy systems.

WRSG – Wounded Rotor Synchronous Generator
PMSG – Permanent Magnet Synchronous Generator
SCIG – Squirrel Cage Induction Generator
DFIG – Doubly Fed Induction Generator
WRIG – Wounded Rotor Induction Generator

2.3.2.1.5 Fixed-Speed Wind Turbines

The frequency of the supply grid, the gear ratio, and the generator design all influence the rotor speed. Fixed-speed wind turbines are typically equipped with an induction generator (squirrel cage or wound rotor) or synchronous generator and are connected to the grid directly. For reactive power assistance in SCIG-based wind turbines, a synchronous condenser or capacitor bank is used. These are designed to run at a certain speed and achieve maximum efficiency. To improve power extraction from the wind, some fixed-speed generators have two winding sets that can be used for low and high speeds.

2.3.2.1.6 Variable-Speed Wind Turbines

Wind turbines with variable speeds capture the most energy from the wind over a wide range of wind speeds. They generate a voltage and frequency output that is adjustable. The generated variable output is converted to a regulated grid

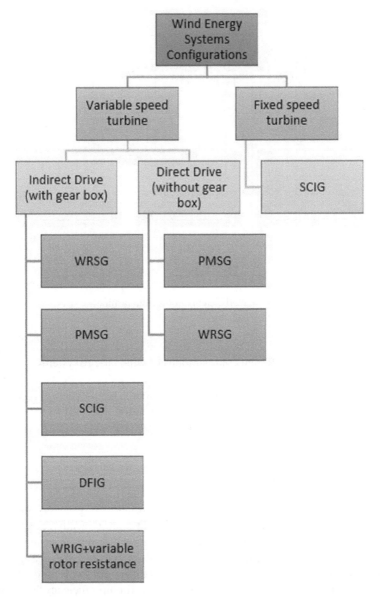

FIGURE 2.16   Wind energy system configurations

voltage and frequency using a power electronic converter. The generator is no longer connected to the grid. Wind turbines with variable speeds are fitted with a doubly fed induction generator, wound rotor, or permanent magnet synchronous generator, and are connected to the grid or off-grid load via a power electronic converter.

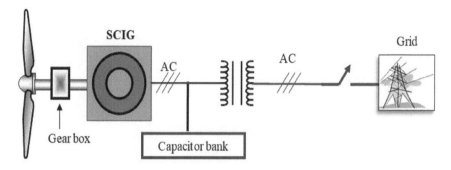

FIGURE 2.17   General structure of a fixed-speed wind turbine

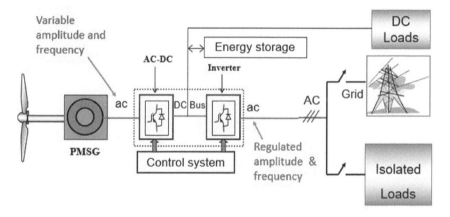

FIGURE 2.18   General structure of a variable-speed wind turbine

2.3.2.2   Grid Integration: Synchronizing with the Grid

The frequency should be as close to the grid frequency as practicable, and preferably a third of a hertz higher. The magnitude of the terminal voltage must match that of the grid, and ideally should be a few percent greater. The two three-phase voltages must have the same phase sequence. Within 5 degrees, the phase angle between the two voltages must be equal.

2.3.2.3   Synchronization Process of Wind Energy Systems

The wind power generator is driven up to speed by utilizing the electrical machine in the motoring mode with the synchronizing breaker open. The machine is switched to generating mode, and the controls are set such that the site and grid voltages match as closely as possible to meet the criteria. As indicated in Figure 2.19, the match is monitored by a synchroscope or three synchronizing lamps, one for each phase. If the lamps remain dark for ¼ to ½ a second, the synchronization breaker is

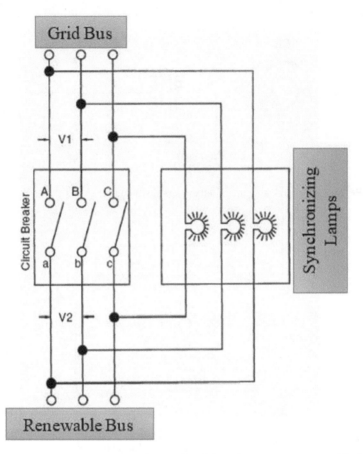

FIGURE 2.19 Synchronizing lamps for synchronizing wind energy systems with the grid

turned off. The difference between the renewable site voltage and the grid voltage at any given time is the voltage across the lamp in each phase.

When the site and grid voltages in all three phases are exactly identical, all three bulbs are dark. It is not enough, however, for the lamps to be dark at any given time. They must remain dark for a long time. Only if the generator and grid voltages have almost the same frequency is this criterion met. If this is not done, one set of the two three-phase voltages will rotate faster than the other, and the phase difference between them will light the lamps.

2.3.3 ENERGY STORAGE SYSTEMS

Electricity is difficult to store on a large scale. Electricity is frequently consumed in the same way that it is produced. Fuel usage in traditional power plants varies with the load requirement on a continual basis. Because wind and solar energy are intermittent sources of energy, they are unable to supply demand at all times, 24 hours a day, 365 days a year. As a result, energy storage is a desirable element to include in such power systems, particularly in standalone facilities.

2.3.3.1 Electrochemical Battery

An electrochemical battery is a type of battery that stores energy in a chemical form. Electrochemical energy exists in a semi-ordered state, somewhere between electrical and thermal energy. Its one-way conversion efficiency is between 85% and 90%.

2.3.3.1.1   Primary Battery

Chemical energy is converted into electric energy by the primary battery. After a full discharge, the electrochemical reaction is irreversible, and the battery is discarded.

2.3.3.1.2   Secondary (Rechargeable) Battery

It is possible to reverse the electrochemical reaction. The battery can be recharged by inserting a direct current from an external source after it has been discharged. In the discharge mode, this turns chemical energy into electric energy, while in the charge mode, it converts electric energy into chemical energy. The efficiency of the round-trip conversion is between 70% and 80%.

To achieve the appropriate battery voltage and current, the battery is made up of multiple electrochemical cells coupled in a series–parallel configuration. The higher the battery voltage, the greater the number of cells in series that is required. At a low electrical potential, typically a few volts, the cell stores electrochemical energy. The capacity of the cell, C, is measured in ampere-hours (Ah), which means that it can deliver C A for one hour or C/n A for n hours. The ampere-hour capacity that the battery can give before the voltage drops below the stipulated limit is stated in terms of the average voltage during discharge. The charge and discharge rates of a battery are expressed in units of its capacity in Ah. The state of charge (*SOC*) of the battery at any time is defined as in equation (2.17):

$$SOC = Ah \ capacity \ remaining \ in \ the \ battery/Rated \ Ah \ capacity \quad (2.17)$$

2.3.3.2 Flywheel

Rotational inertia stores kinetic energy in a flywheel. This energy can be efficiently transferred from and to electricity. The round-trip conversion efficiency of a big flywheel system can approach 90%, which is significantly higher than that of a battery.

2.3.3.2.1   Flywheel System Components

Figure 2.20 depicts the general structure of a flywheel. A robust hub connects a high-speed rotor to the shaft. In the high-speed rotor, a flywheel incorporates bearings

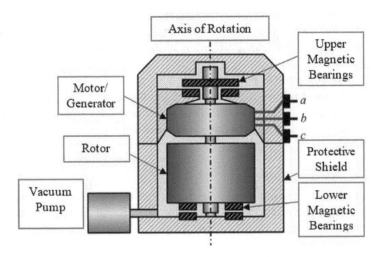

FIGURE 2.20   General components of a flywheel

with a good lubrication system or magnetic suspension. An electromechanical energy converter is a mechanism that can charge and discharge energy by acting as both a motor and a generator. To drive the motor and condition the generating power, power electronics are provided. The magnetic bearings and other operations can be controlled using control electronics.

2.3.3.2.2   Benefits of a Flywheel over a Battery
The following are some of the advantages of a flywheel versus a battery:

- A significant amount of energy can be stored per unit of weight and volume
- High depth of discharge (DoD)
- High peak-power capabilities without overheating problems
- Design flexibility for a given voltage and current
- Better power quality since the electrical machine is more rigid than the battery
- Simple power management, as the SOC is determined solely by the speed.

3 Overview of Microgrids

3.1 WHAT IS A MICROGRID?

A microgrid is an electrical system comprised of distributed energy resources and loads that operates in parallel to the utility grid or as an isolated system. A microgrid can be defined by three key characteristics, as follows.

- **Local**
  A microgrid is focused on catering for nearby customers. Therefore, the energy produced is referred to as local energy. This characteristic distinguishes a microgrid from the utility grid. Conventional utility grids are based on centralized generation that involves long distance transmission and distribution over the power network. This results in losses of generated electrical energy. Microgrids overcome this issue by generating at or near the point of consumption.
- **Independent**
  Grid-connected solar PV systems that are installed and owned by consumers for domestic or commercial purposes are restricted to generating electricity when the utility grid is unavailable. Unlike a solar PV system, a microgrid maintains an uninterrupted power supply catering for the local loads by operating as an electrical island even when the utility grid has failed. The multimode operation (grid-connected and islanded) makes a microgrid an independent electrical island. However, there are certain differences between the islanded operation and the grid-connected mode in terms of control, balancing demand and supply, prioritizing and catering for loads, etc. Novel control, protection and energy management methods for microgrids ensure a seamless transition between multiple modes of operation, thus making a microgrid independent.
- **Intelligent**
  Most advanced microgrids are intelligent. There are novel intelligent control and energy management approaches for microgrids that satisfy the goals of the microgrid as well as those of consumers, in terms of maximizing the utilization of renewable generation, managing the demand side, minimizing

DOI: 10.1201/9781003216292-3

the cost of energy etc. For example, an advanced controller can track real-time changes in the power prices on the central grid. (Wholesale electricity prices fluctuate constantly based on electricity supply and demand.) If energy prices are inexpensive at any point, it may choose to buy power from the central grid to serve its customers, rather than use energy from, say, its own solar panels. Its own solar panels could instead charge its battery systems. Later in the day, when grid power becomes expensive, the microgrid may discharge its batteries rather than use grid power.

3.2 MICROGRID POWER ARCHITECTURE

3.2.1 MICROGRID STRUCTURE AND COMPONENTS

A microgrid is a small-scale electrical system that is designed to provide power for a small community. As given in Figure 3.1, a microgrid is comprised of distributed energy resources such as solar photovoltaic, wind turbines, and fuel cells, distributed loads, power electronic interface units (DC/DC, DC/AC, AC/DC and AC/AC converters), and a point of common coupling (PCC) [1–3].

Large rotating machines introduce inertia to the system, such that there will be an excess time for the system to stabilize. Since microgrids are comprised of renewable sources which lack inertia, an energy storage unit has become a necessity. A battery backup or other energy storage unit in a microgrid functions as a synchronous generator. Microgrids often have technologies such as solar PV or micro turbines, both of which require power electronic interfaces such as DC/AC or DC/AC/DC converters to get connected to the electrical system.

Microgrids operate in either grid-connected or islanded mode. When the utility grid fails, the PCC allows the microgrid to disconnect from the grid and operate as an independent, electrical island.

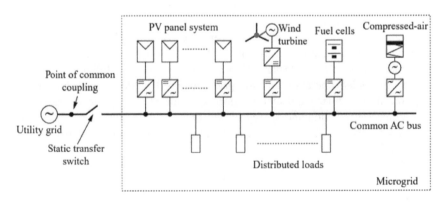

FIGURE 3.1   General structure of a microgrid

3.2.2 Types of Power Architecture

Microgrid power architecture can be classified based on the available power lines, the nature of the sources, and the operation, as shown in Figure 3.2.

- According to type of power lines
  - AC power architecture
  - DC power architecture
  - DC/AC power architecture
- According to nature of sources
  - Solar—battery based power architecture
  - Solar–wind based power architecture
  - Solar–diesel based power generator
  - Solar–mini hydro based power
- According to operation
  - Reconfigurable architecture

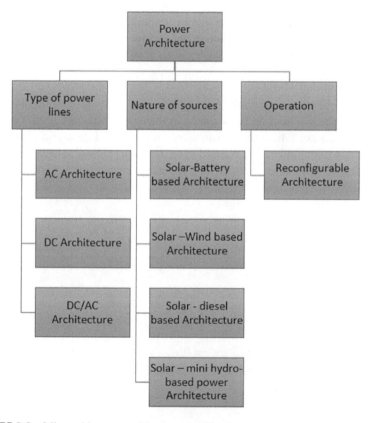

FIGURE 3.2 Microgrid power architecture classification

3.3   OPERATION OF MICROGRID

3.3.1   MODES OF OPERATION

A microgrid is generally operated while connected to the main grid when this is available. In a case of grid outage, it will shift to operating in islanded mode, using its own generation sources, such as renewable sources like solar panels, batteries, diesel generators, and so on.

3.3.1.1   Grid-Connected Mode

In the grid-connected mode, the utility grid is in the active state. The static switch is kept closed as shown in Figure 3.3. The utility regulates the frequency and voltage of the microgrid such that the distributed generating units follow the main grid's frequency variations.

In grid-connected microgrids, the demand–supply mismatch is overcome by injecting the excess power generated in the microgrid to the utility grid, and consuming grid power when the in-house generation is insufficient to cater for the local loads. The energy management functions of grid-connected microgrids include minimization of the cost of energy, minimization of the grid dependency of the microgrid, optimum utilization of the energy storage, generation scheduling and dispatching of DERs, and strategic and economic operations.

3.3.1.2   Islanded Mode

In the islanded mode, the utility grid does not supply power to the microgrid. The static switch is kept open, as shown in Figure 3.4. The microgrid frequency and voltage do not follow the variations of the main grid. In a grid outage or as scheduled, a microgrid can be isolated from the main grid's distribution system at the PCC.

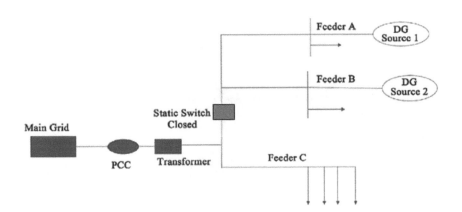

FIGURE 3.3   Grid-connected microgrid

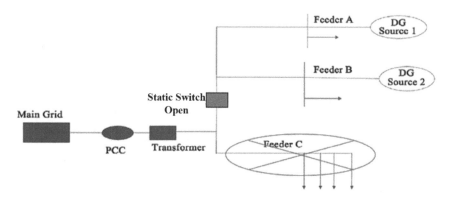

FIGURE 3.4 Islanded microgrid

As shown in Figure 3.4, all the feeders, except for feeder C, are supplied by the distributed energy resources of the microgrid. However, with the limited generation capacity and reliance on intermittent energy resources, it is necessary to minimize the demand–supply deficit and maintain an uninterrupted power supply to the critical loads. A priority order for catering for loads has to be set up to ensure that important loads get an uninterrupted power supply. Feeder C is considered as a feeder with non-critical loads such that, during the islanded mode of operation, it is not supplied by the micro sources.

3.3.2 DEMAND–SUPPLY BALANCE

Demand–supply balance determines the frequency stability of a power system. Figure 3.5 is a graphical representation of the effect of demand–supply variation on the system frequency through the operation of a balance scale. A standard frequency is essential to avoid damage to equipment resulting from multiple frequencies operating alongside each other. When it comes to providing electricity on a national basis, this has huge implications.

It is more important to maintain frequency stability across the power system than to maintain an exact standard frequency. For example, the standard frequency in the United State is 60Hz while it is 50Hz in Great Britain. Japan follows multiple standard frequencies, such that the western and eastern regions of the country run at 60Hz and 50 Hz respectively. Stepping up and down the frequency of the electricity that flows between the two regions is handled by power stations located across the middle of the country.

When the demand exceeds the generated power, the rotational generators may start to decelerate, resulting in a decrease in system frequency. In such under-frequency conditions a load shedding plan is activated in order to avoid power cuts. This is because, if the frequency falls too much, the power plants switch off one after another, until there is a complete collapse of the grid (that is, a power

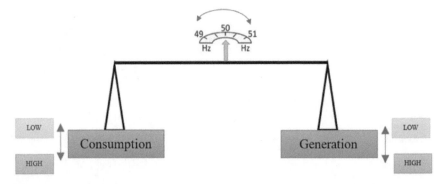

FIGURE 3.5 Frequency deviation from the nominal value represents mismatch between active power generation and consumption

blackout). When the generation exceeds the demand, the rotational generators may start to accelerate, resulting in an increase in system frequency.

Unlike a conventional power system, a microgrid with renewable energy resources such as solar PV lacks inertial generation, leading to very rapid frequency and voltage variations within the microgrid in a demand–supply mismatch.

3.3.3 Types of Distributed Generators Based on Different Operating Conditions

3.3.3.1 Grid-Forming Units

When a microgrid operates in the islanded mode, the system voltage and frequency should be regulated to avoid demand–supply mismatches. This is achieved through grid-forming units. As the name implies, in the absence of the main grid, a grid-forming unit regulates the voltage and frequency in a similar manner to the regulation in the grid-connected operation. In the grid-connected operation, the voltage and frequency regulation is done through the main grid, converting grid-forming units to grid-feeding units. Therefore grid-forming units should be designed and controlled to operate in both islanded and grid-connected modes accordingly.

Examples of grid-forming units are gas turbine generators and CHP systems, diesel generators, and energy storage.

3.3.3.2 Grid-Feeding Units

Non-dispatchable distributed generators such as solar PV and wind generators operate as grid-feeding generators by feeding generated power regardless of the amount or as suggested by the microgrid operator. Grid-feeding units operate under the assumption that the microgrid is readily available to accept their generation.

Examples of grid-feeding units are wind units, solar PV units, and fuel cells.

3.3.3.3 Grid-Following Units

Grid-following distributed generators partially cater for local loads by following the microgrid frequency and terminal voltages. Dispatchable distributed generators such as battery energy storage and micro turbines that have sufficient capacity and fast-responding abilities can transfer from the grid-following mode to the grid-forming mode accordingly. Generally, distributed generators that are assigned to be grid-following units assist the grid-forming generators by sharing loads with a droop characteristic.

3.3.4 Types of Electrical Load

A device or electrical component that consumes electrical energy and converts it to another kind of energy is known as an electrical load. Electrical loads include equipment like lamps, air conditioners, motors, and resistors, to name but a few. In other words, an electrical load is a section of a circuit that connects to a distinct output terminal.

A common categorization of electrical loads considers their nature as resistive, capacitive or inductive loads, or a combination of these. In an alternating current (AC) configuration, these consume power differently. Lighting, mechanical, and thermal loads are represented by capacitive, inductive, and resistive load types.

3.3.4.1 Resistive Loads

Incandescent lights and electric heaters are two common examples of AC resistive loads. Electrical power is consumed by resistive loads in such a way that the current wave remains in phase with the voltage wave and the power factor is unity.

3.3.4.2 Capacitive Loads

Current and voltage are out of phase in a capacitive load. The difference is that, with a capacitive load, the current reaches its maximum value before the voltage reaches its maximum value, such that the current waveform precedes the voltage waveform. Even though certain electrical loads are classified as resistive or inductive, there are no purely capacitive loads. Capacitive loads are frequently used in electrical substations to improve the system's overall "power factor".

3.3.4.3 Inductive Loads

In an inductive load, the current wave lags behind the voltage wave, such that the peak of the sinusoidal waveform of the current appears after the peak of the voltage waveform. As a result, the power factor is lagging. Transformers, motors, and coils are examples of inductive loads.

Current and voltage waveforms and phasor diagrams for resistive, inductive and capacitive loads are given in Figure 3.6.

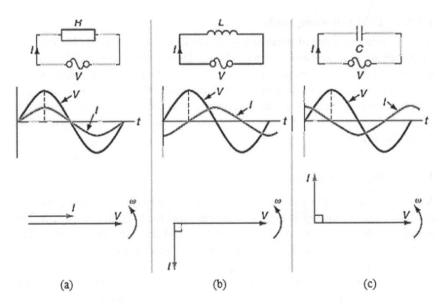

FIGURE 3.6 Current and voltage waveforms and phasor diagrams for (a) resistive load, (b) inductive load and (c) capacitive load

3.3.4.4 Combination Loads

The majority of loads are not purely resistive, capacitive, or inductive. Various combinations of resistors, capacitors, and inductors are used in many real loads. Such loads have power factors less than unity and are either lagging or leading.

The classification of electrical loads is based on various factors, as shown in Figure 3.7.

3.4 TYPES OF MICROGRID CONTROL ARCHITECTURE

3.4.1 CENTRALIZED CONTROL

In centralized control, a central control unit is available to which all the other control units are connected. Generally, the central control unit is responsible for data storing. Other control units have access to the data stored within the central control unit. The central controller is also referred to as a central server. Setting up a centralized control system is relatively easy because of its simpler architecture. These systems can be easily maintained.

However, there are certain limitations associated with centralized control systems. If there is a failure of the central server or controller, the other connected control units cannot access data. In a centralized control system, other connected units depend on the availability of the central unit. This challenges the consistent performance of the control system by making it prone to single point failure. With regard to the security and privacy of the other control units, there are also limitations introduced by a centralized control architecture [4].

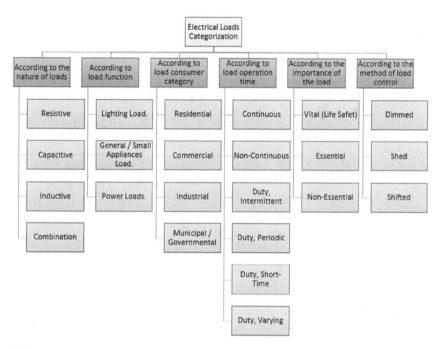

FIGURE 3.7 Electrical load categorization

3.4.2 DECENTRALIZED CONTROL

Unlike centralized systems, decentralized systems do not have a central control unit. Instead of a single central server unit, there are multiple such units with data storing and retrieving capabilities. The availability of multiple data access points means that the overall function of the control system is much more efficient in comparison to centralized systems. Decentralized control units remain in active status until all the server units fail. Overall, a server failure in a decentralized system may limit some of the control functions, but the performance is still advanced in comparison to centralized systems [5].

However, decentralized systems are also prone to failures if they are not designed to maintain the functions after a failure of one or more server units. Several data storing units may result in higher maintenance costs.

3.4.3 DISTRIBUTED CONTROL

Distributed systems are similar to decentralized systems as these systems do not depend on a single central control unit. However, distributed control systems enable equal capabilities and data accessing for all the control units. The significance of distributed systems is the allocation of software and hardware facilities among all the control units to mitigate the functional effects of component failures. Distributed systems are flexible for modifications and expansions. Like centralized and decentralized systems,

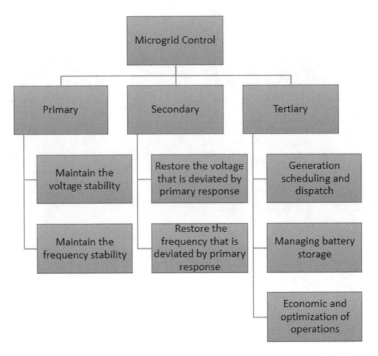

FIGURE 3.8 Microgrid control hierarchy

distributed systems suffer from data security and privacy concerns. In addition, resource allocation may result in higher maintenance costs [6, 7].

3.4.4 HIERARCHICAL CONTROL

Microgrids can essentially be controlled in the same way as a main grid by using a three-level hierarchical control. The hierarchical control of a microgrid is presented in Figure 3.8. The three control levels are defined as primary, secondary, and tertiary, based on their speeds of response, operational timeframe and other infrastructural requirements [8]. Primary control involves quick responses based on local measurements without communicating with other elements, and is based on droop control. For example, the detection of islanding and power balance is the function of primary control. As a result of the primary response, the rated frequency and voltage deviate from the nominal values. Secondary control is responsible for restoring the rated frequency and voltage of the microgrid. Tertiary control involves energy management functions of the microgrid such as the economic dispatch of DERs, grid consumption minimization, the optimum utilization of energy storage units, and so on.

3.4.4.1 Droop Control

Droop control is a prominent method of primary control that adjusts the output active power of the controlled asset to regulate frequency. Droop control is the key

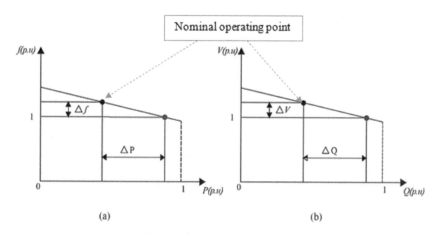

FIGURE 3.9 Droop characteristic curves: (a) frequency–active power (F-P) characteristic curve; (b) voltage–reactive power (V-Q) characteristic curve

to sharing demand across generators in autonomous microgrids where the utility grid is unavailable. Voltage source converters (VSCs) are frequently used in the islanded operating mode of microgrids. The microgrid's voltage and frequency are managed by local control loops. Control techniques based on droop characteristics are commonly used to avoid circulating currents between parallel inverters connected to the microgrid.

The concept of proportional power sharing between synchronous generators in complex interconnected power systems led to the development of droop control approaches. There are two main droop control methods: frequency–active power (F-P) and voltage–reactive power (V-Q). The characteristics of F-P and V-Q droop controls are given in Figure 3.9.

An imbalance between a generator's input mechanical power and its output electric active power creates a change in the rotor speed, which results in a frequency deviation. Similarly, variations in output reactive power cause the voltage magnitude to vary. The frequency–power droop control method is defined for conventional distributed generator units like synchronous generators, and it can be recreated in electronically interfaced distributed generating units as well.

3.4.4.2 Primary Control

Generally, primary control is based on responding promptly and relying solely on local measurements without communicating further with other control elements.

Microgrids rely heavily on power sharing. In general, power sharing implies sharing the load among paralleled distributed generating units in the microgrid, as well as sharing the load with the main grid by controlling the microgrid–main grid power exchange. Therefore, power sharing plays an important role at the primary control level. In addition, primary control is involved in islanding detection, controlling the power output, and so on.

3.4.4.3 Secondary Control

When disturbances occur as a result of the primary response, the frequency and voltage of the microgrid may change from their nominal values. Secondary control is responsible for restoring the rated frequency and voltage in the microgrid after such a primary control response in hierarchical control.

3.4.4.4 Tertiary Control

Tertiary control ensures a microgrid's reliable, secure, and cost-effective operation, in both grid-connected and islanded modes. The energy management functions of a microgrid are optimized by the tertiary control, by taking the interaction between the microgrid and the utility grid into account. For example, operating the microgrid as an electrical island, resynchronization, and grid dependency of the microgrid involving power exchange between the microgrid and the main grid are some of the functions that are optimized through the tertiary response in hierarchical control.

Figure 3.10 shows how the nominal operating point of a generator varies with hierarchical control following primary, secondary and tertiary responses. Let's consider an islanded microgrid. The starting operating point is at the nominal voltage and frequency (1). When a deficit in power generation occurs as a result of a sudden increase in the demand for locally supplied real power and a change in reactive power, the grid-forming generator should adjust its contribution to cater for the increase in real power. This results in a decrease of the operating frequency. This initial change of the operating point from 1 to 2 is due to the primary response. At the second operating point (2), the operating frequency has deviated from the nominal frequency. The secondary control restores the operating frequency to the nominal value, as in Figure 3.10, by changing the operating point from 2 to 3. As a result, the operation of the generator is shifted to a new droop characteristic curve. The tertiary response is an optional control based on the energy management function optimization of the microgrid and changes the operating point from 3 to 4.

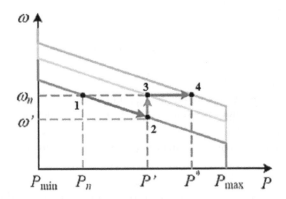

FIGURE 3.10 Primary, secondary, and tertiary control responses of hierarchical control

3.5   ADVANTAGES AND DISADVANTAGES OF MICROGRIDS

3.5.1   ADVANTAGES OF MICROGRIDS

The microgrid concept can be adapted as a solution for the current energy crisis. The efficiency of a microgrid is higher than that of a conventional grid. A microgrid comprises micro sources that are located close to the point of consumption, such that the transmission losses are significantly lower than with a conventional grid. In addition, microgrids encourage the high penetration of renewable energy resources (RERs) by providing suitable integrating platforms for loads and distributed generators.

The contribution of microgrids in reducing carbon dioxide emissions and avoiding the consequences of large-scale land use has major environmental benefits. One of the main advantages of a microgrid is its ability to operate in scenarios when the grid is available and when it has failed. The seamless transition from grid-connected to islanded mode and vice versa ensures an uninterrupted power supply to the loads within the electrical boundary of the microgrid. When the power system needs this, microgrids can isolate themselves from the grid to reduce the load on the grid with no, or minor, interruption to the loads [9–10].

However, microgrids can cater for critical loads with a reliable and good quality power supply. Microgrid energy management system is advantageous in terms of reducing the cost of electricity to the users, the optimum utilization of renewable generation, active participation in demand response programs, and so on. When the generation of distributed energy resources is limited, non-critical microgrid loads are curtailed accordingly to ensure an uninterrupted power supply to crucial loads. In a novel approach, closely located islanded microgrids can be networked as microgrid clusters that are physically connected and functionally interoperable. The networking of microgrids allows islanded microgrids to share extra generation resource capacities to minimize demand–supply deficits among themselves and enhance the resilience and reliability of the power system [11–13].

3.5.2   DISADVANTAGES OF MICROGRIDS

However, there are several limitations related to microgrids. The RERs in microgrids lack inertia (rotational kinetic energy), spinning reserves and storage. Therefore, these three attributes should be artificially created for RERs. The inertia can be artificially created by adding energy storage units. Energy storage units can absorb mismatches in demand and supply within the microgrid and help in stabilizing the microgrid after a disturbance. The continuous control and monitoring of intermittent RERs is another challenge when implementing microgrids. Mainly in islanded mode, voltage, frequency and power quality should be controlled such that these parameters comply with the standards in addition to the demand–supply balance.

3.6   NETWORKED MICROGRIDS

Nowadays consumers' high reliance on electricity for daily activities and critical services has created concerns about power system reliability for low-impact high-frequency events and, equally, resilience for high-impact low-frequency events. The vulnerability of the power infrastructure to high-impact low-frequency events such as extreme weather, earthquakes, tsunamis, man-made outages, and so on can have huge socio-economic costs in terms of power interruptions and financial losses [14]. The resilience of the power system refers to adequate preparation for, thorough response to, and rapid recovery from significant disturbances due to extreme events. However, the power system resilience can be considered as the resilience of the distribution network as it is highly vulnerable to catastrophic events [15].

A conventional power system is designed by following N-1 or N-2 contingency criteria to ensure the system is reliable and can eliminate the possibility that a failure of one or two components results in cascading failures of the network. Even though reliability improvement measures are taken, power systems are still vulnerable to catastrophic events. Therefore, power system planners are concerned with modifying the existing power infrastructure by taking strengthening measures such as using underground cables, improving the strength of towers, etc. However, these infrastructure modifications alone cannot manage the adverse impacts of extreme events [16].

Together with the distributed generation concept, microgrids have come into the picture as localized distribution networks that consist of distributed energy resources such as wind energy, solar energy, storage systems, electric vehicles, and others [17]. As discussed previously, microgrids operate in two different modes: grid-connected (operating in parallel to the main grid) and islanded (disconnected from the main grid while supplying customers only within their electric boundary). In the islanded mode, the voltage and the frequency regulation of the microgrid are carried out through the grid-forming distributed energy resources. The demand–supply balance within the electric boundary of the microgrid is maintained by utilizing renewable energy resources such as solar PV and wind, diesel generators and energy storage units.

There is ample research dedicated to individual microgrid control and operation to enhance the power system resilience. An approach for resilience improvement for microgrids based on self-healing is proposed in [18]. In [19] and [20], respectively, the resource allocation for forming a microgrid for a resilient electric grid, and the proactive management of microgrids to improve resilience for windstorms, are presented. Resilience improvement of power systems using networked microgrids has recently become an emerging research area. Therefore, several studies have been conducted in this area, such as on the design of resilient, large-scale distribution feeders based on networked microgrids [21], resilient power sharing strategies in networked microgrids [11], and the enhancement of the resilience response by peer-to-peer energy bartering in networked microgrids [12].

3.7   EXAMPLE: MICROGRID MODELING AND SIMULATION

This example depicts the performance of a simplified model of a small-scale microgrid during a typical day (24 hours). A single-phase AC network makes up the microgrid. The utility grid, a solar power generation system, and a battery are the available energy sources. The simulation model for the microgrid modeling is given in Figure 3.11.

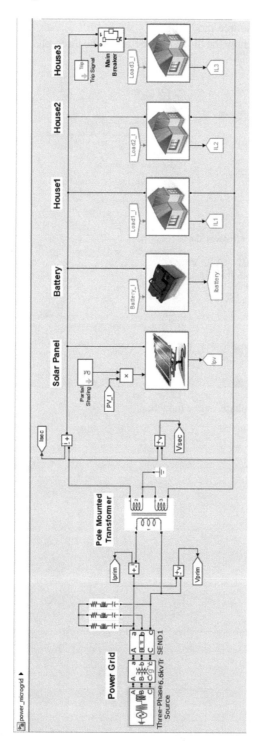

FIGURE 3.11 Microgrid simulation model in MATLAB simulink

Step 1: Three-phase source modeling

Figure 3.12 shows how the block parameters of the three-phase source are set up.

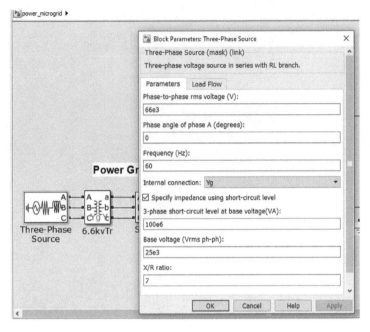

FIGURE 3.12 Three-phase source modeling

Step 2: Three-phase transformer modeling

Figure 3.13 shows how the block parameters of the three-phase transformer are set up.

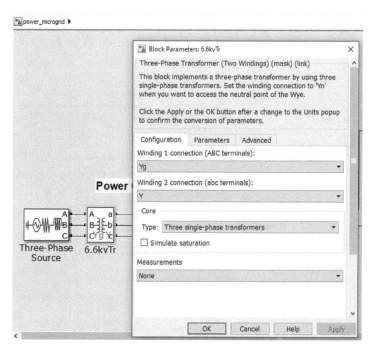

FIGURE 3.13   Three-phase transformer modeling

Step 3: Three-phase PI section line

The three-phase PI section line block implements a balanced three-phase transmission line model with parameters lumped in a PI section. This model consists of one set of RL series elements (a series connection of a resistor (R) and an inductor (L)) connected between the input and the output terminal, and two sets of shunt capacitances lumped at both ends of the line. Figure 3.14 shows how the block parameters of the three-phase PI section line are set up.

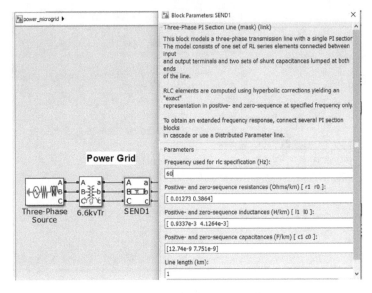

FIGURE 3.14   Three-phase PI section line modeling

Step 4: Series RLC load

A series RLC branch represents a series connection of a resistor, an inductor and a capacitor. A series RLC branch can be either single-phase or three-phase. The user can specify which elements are in the branch by changing the branch type. Figure 3.15 shows how the block parameters of the series RLC load are set up. There are three series RLC loads that are connected in parallel.

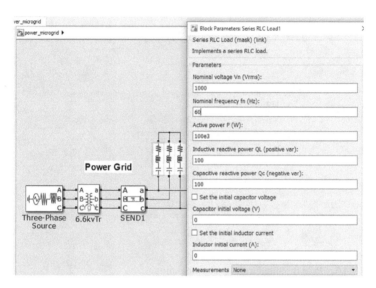

FIGURE 3.15   Series RLC load modeling

Step 5: Pole mounted transformer modeling

A transformer mounted on a pole connects the micro-array to the utility grid, stepping down the voltage from 6.6 kV to 200 V. Figure 3.16 shows how the block parameters of the three winding linear transformers are set up.

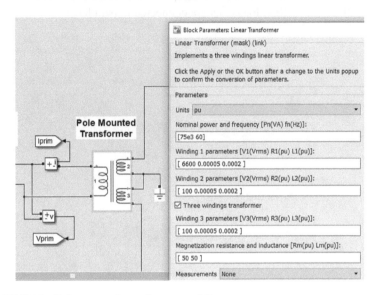

FIGURE 3.16   Pole mounted transformer modeling

Step 6: Microgrid elements modeling

As shown in Figure 3.17, solar panels and batteries are both DC power sources that are converted to single-phase AC. The control technique assumes that the device does not rely completely on power from the grid, and that the electricity generated by the solar panels and stored in the batteries is always sufficient. Three regular residences utilize energy (maximum of 2.5 kW).

A battery controller is in charge of the battery, as shown in Figure 3.18. When there is excess energy in the micro-network, the battery absorbs it, and when there is a power deficit in the micro-network, the battery delivers additional power.

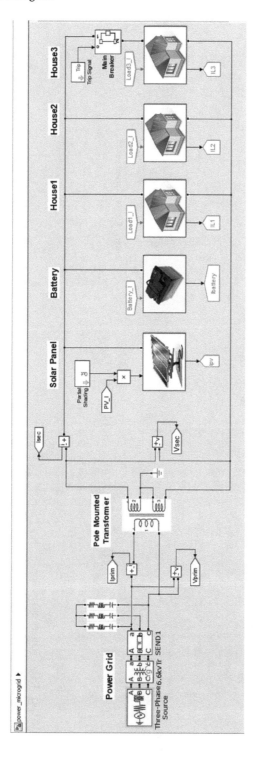

FIGURE 3.17 Microgrid elements modeling

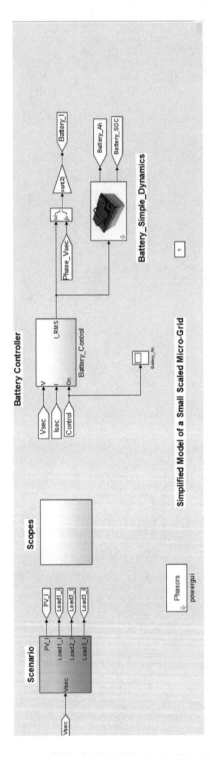

FIGURE 3.18 Battery controller

- Solar PV array modeling

The solar PV subsystem can be modeled as shown in Figure 3.19 using MATLAB Simulink.

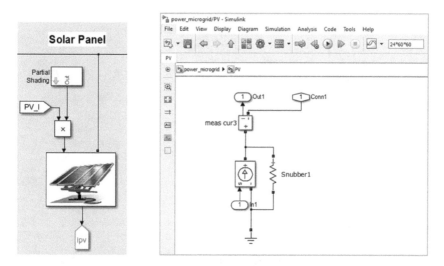

FIGURE 3.19 Solar PV subsystem

The parameters of the snubber and the current source can be set as given in Figures 3.20 and 3.21 respectively.

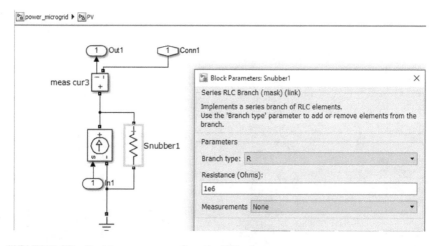

FIGURE 3.20 Snubber parameters for solar PV subsystem

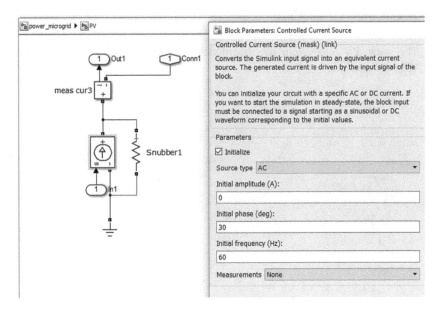

FIGURE 3.21   Current source parameters for solar PV subsystem

To demonstrate the performance of the solar array in a practical scenario, a partial shading block is added to the solar PV system. Figure 3.22 shows how the parameters of the partial shading block are set up. Here, within the specified time range defined by the start and duration parameters, the output is set to "Factor". Outside the specified range, the output is kept at "Factor = 1".

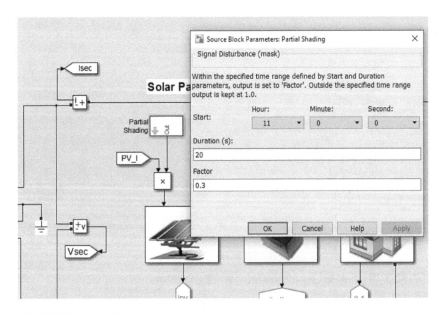

FIGURE 3.22   Partial shading block

• Battery backup modeling

The battery subsystem can be modeled as given in Figure 3.23 using MATLAB Simulink. The snubber and current source modeling of the battery subsystem is similar to the solar PV system modeling.

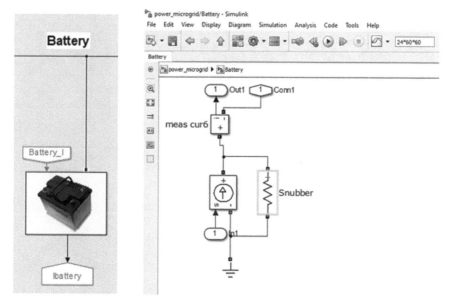

FIGURE 3.23   Battery backup modeling

- Residential load modeling

The residential load subsystem can be modeled as given in Figure 3.24 using MATLAB Simulink. Here three regular residences utilize energy (maximum of 2.5 kW).

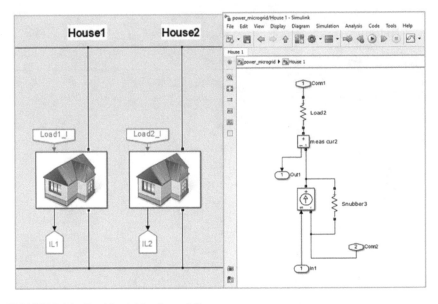

FIGURE 3.24   Residential loads modeling

The parameters of the load, snubber and the current source can be set as given in Figures 3.25, 3.26 and 3.27 respectively.

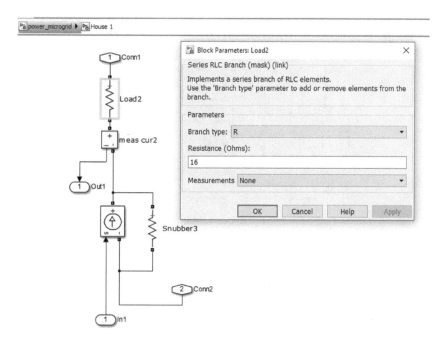

FIGURE 3.25 Load parameters

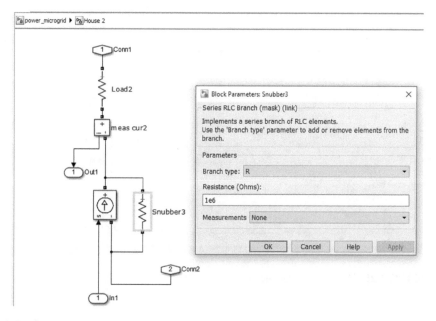

FIGURE 3.26 Snubber parameters for residential loads modeling

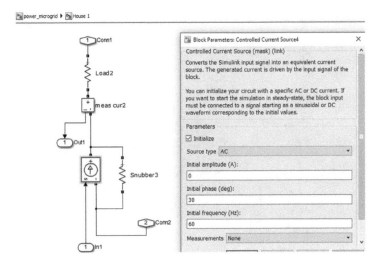

FIGURE 3.27   Current source parameters for residential loads modeling

- Main breaker

Figure 3.28 shows how the block parameters of the main breaker are set up.

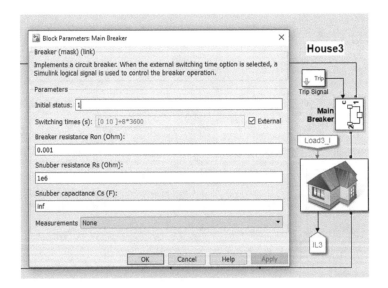

FIGURE 3.28   Main breaker modeling

Figure 3.29 shows how the trip signal is generated for the main circuit breaker by specifying the fault timing and fault duration.

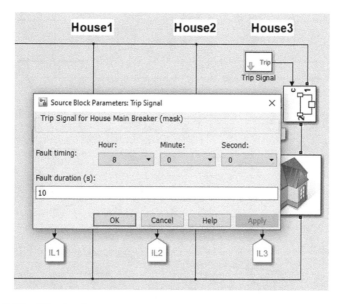

FIGURE 3.29 Trip signal parameters

Simulation results and discussion

The simulation results generated from the above Simulink model are given in Figure 3.30.

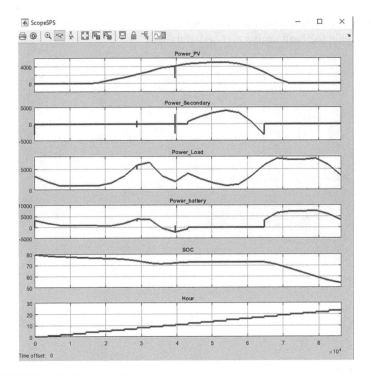

FIGURE 3.30   Simulation results

From 20h to 4h, the solar power generated is 0 W. The solar power generated reaches its peak amount (5 kW) from 14h to 15h.

Like the typical load change in ordinary houses, the electric power load reaches peak consumption at 9h (6,500 W), 19h, and 22h (7,500 W).

From 0h to 12h and from 18h to 24h, battery control is performed by the battery controller. The battery control performs tracking control of the current so that the active power which flows into the system power from the secondary side of the pole transformer is set to 0. Then, the active power of the secondary side of the pole mounted transformer is always around zero.

The storage battery supplies the deficit current when the power of the microgrid is insufficient, and absorbs surplus current from the microgrid when its power surpasses the electric load.

From 12h to 18h, battery control is not performed. The SOC (State of Charge) of the storage battery is fixed at a constant and does not change, since neither charge nor discharge of the battery is performed by the battery controller. When there is a power shortage in the microgrid, the system power supplies the deficit power. When there is surplus power in the microgrid, surplus power is returned to the system.

At 8h, the electricity load of an ordinary house No. 3 is set to OFF for 10 seconds by the breaker. A spike is observed in the active power on the secondary side of the pole transformer and the electric power of the storage battery.

REFERENCES

[1] A. Hirsch, Y. Parag, and J. Guerrero, "Microgrids: A review of technologies, key drivers, and outstanding issues," *Renew. Sustain. Energy Rev.*, vol. 90, pp. 402–411, 2018, doi: 10.1016/j.rser.2018.03.040.

[2] M. F. Zia, E. Elbouchikhi, and M. Benbouzid, "Microgrids energy management systems: A critical review on methods, solutions, and prospects," *Appl. Energy*, vol. 222, pp. 1033–1055, 2018, doi: 10.1016/j.apenergy.2018.04.103.

[3] J. Kumar, A. Agarwal, and V. Agarwal, "A review on overall control of DC microgrids," *J. Energy Storage*, vol. 21, pp. 113–138, 2019, doi: 10.1016/j.est.2018.11.013.

[4] D. Espín-Sarzosa, R. Palma-Behnke, and O. Núñez-Mata, "Energy management systems for microgrids: Main existing trends in centralized control architectures," *Energies*, vol. 13, no. 3, p. 547, 2020, doi: 10.3390/en13030547.

[5] A. Bani-Ahmed, M. Rashidi, A. Nasiri, and H. Hosseini, "Reliability analysis of a decentralized microgrid control architecture," *IEEE Trans. Smart Grid*, vol. 10, no. 4, pp. 3910–3918, 2019, doi: 10.1109/TSG.2018.2843527.

[6] P. Lin, P. Wang, J. Xiao, C. Jin, and K. L. Hai, "A distributed control architecture for hybrid AC/DC microgrid economic operation," *Proc. 13th IEEE Conf. Ind. Electron. Appl. ICIEA 2018*, pp. 690–694, 2018, doi: 10.1109/ICIEA.2018.8397802.

[7] Q. Zhou, M. Shahidehpour, A. Paaso, S. Bahramirad, A. Alabdulwahab, and A. Abusorrah, "Distributed control and communication strategies in networked microgrids," *IEEE Communications Surveys & Tutorials*, vol. 22, no. 4, pp. 2586–2633, 2020, doi: 10.1109/COMST.2020.3023963.

[8] D. Y. Yamashita, I. Vechiu, and J. P. Gaubert, "A review of hierarchical control for building microgrids," *Renew. Sustain. Energy Rev.*, vol. 118, p. 109523, 2020, doi: 10.1016/j.rser.2019.109523.

[9] M. Hamidi, O. Bouattane, and A. Raihani, "Microgrid energy management system: Technologies and architectures review," *Proc. 2020 IEEE Int. Conf. Moroccan Geomatics, MORGEO 2020*, 2020, doi: 10.1109/Morgeo49228.2020.9121885.

[10] S. Pradhan, D. Mishra, and M. K. Maharana, "Energy management system for micro grid pertaining to renewable energy sources: A review," *2017 International Conference on Innovative Mechanisms for Industry Applications (ICIMIA)*, pp. 18–23, 2017, doi: 10.1109/ICIMIA.2017.7975612.

[11] L. Ren *et al.*, "Enabling resilient distributed power sharing in networked microgrids through software defined networking," *Appl. Energy*, vol. 210, pp. 1251–1265, 2018, doi: 10.1016/j.apenergy.2017.06.006.

[12] M. Mehri Arsoon and S. M. Moghaddas-Tafreshi, "Peer-to-peer energy bartering for the resilience response enhancement of networked microgrids," *Appl. Energy*, vol. 261, p. 114413, 2020, doi: 10.1016/j.apenergy.2019.114413.

[13] B. Chen, J. Wang, X. Lu, C. Chen, and S. Zhao, "Networked microgrids for grid resilience, robustness, and efficiency: A review," *IEEE Trans. Smart Grid*, vol. 12, no. 1, pp. 18–32, 2021, doi: 10.1109/TSG.2020.3010570.

[14] E. Galvan, P. Mandal, and Y. Sang, "Networked microgrids with roof-top solar PV and battery energy storage to improve distribution grids resilience to natural disasters," *Int. J. Electr. Power Energy Syst.*, vol. 123, p. 106239, 2020, doi: 10.1016/j.ijepes.2020.106239.

[15] H. Raoufi, V. Vahidinasab, and K. Mehran, "Power systems resilience metrics: A comprehensive review of challenges and outlook," *Sustain.*, vol. 12, no. 22, p. 9698, 2020, doi: 10.3390/su12229698.

[16] Z. Li, M. Shahidehpour, F. Aminifar, A. Alabdulwahab, and Y. Al-Turki, "Networked microgrids for enhancing the power system resilience," *Proc. IEEE*, vol. 105, no. 7, pp. 1289–1310, 2017, doi: 10.1109/JPROC.2017.2685558.

[17] A. Cagnano, E. De Tuglie, and P. Mancarella, "Microgrids: Overview and guidelines for practical implementations and operation," *Appl. Energy*, vol. 258, p. 114039, 2020, doi: 10.1016/j.apenergy.2019.114039.

[18] K. Khalili, I. Pourkeivani, and M. Abedi, "Micro-grids resilience enhancement using self-healing capability," *34th Int. Power Syst. Conf. PSC 2019*, pp. 750–756, 2019, doi: 10.1109/PSC49016.2019.9081482.

[19] K. S. A. Sedzro, A. J. Lamadrid, and L. F. Zuluaga, "Allocation of resources using a microgrid formation approach for resilient electric grids," *IEEE Trans. Power Syst.*, vol. 33, no. 3, pp. 2633–2643, 2018, doi: 10.1109/TPWRS.2017.2746622.

[20] M. H. Amirioun, F. Aminifar, and H. Lesani, "Resilience-oriented proactive management of microgrids against windstorms," *IEEE Trans. Power Syst.*, vol. 33, no. 4, pp. 4275–4284, 2018, doi: 10.1109/TPWRS.2017.2765600.

[21] A. Barnes, H. Nagarajan, E. Yamangil, R. Bent, and S. Backhaus, "Resilient design of large-scale distribution feeders with networked microgrids," *Electr. Power Syst. Res.*, vol. 171, pp. 150–157, 2019, doi: 10.1016/j.epsr.2019.02.012.

4 Novel Approaches to Microgrid Functions

4.1 RECONFIGURABLE POWER ELECTRONIC INTERFACES

4.1.1 INTRODUCTION TO POWER ELECTRONIC INTERFACES

Power electronic interfaces improve the quality and value of electricity, from generation to utilization. Commonly, power electronic interfaces are used when integrating distributed energy resources to microgrids. The power electronic interface does not store electrical energy but converts the power received from the distributed energy resources to a desired voltage and frequency. Some of the important functions of power electronic interfaces can be presented as follows.

Integration requirements: When multiple distributed energy resources are available in a microgrid, the different characteristics of the components mean that power conversion (AC to DC, DC to AC, DC to DC, AC to AC) is required. Depending on the architecture of the microgrid as AC or DC or hybrid, the power electronic interface units required may vary. In addition to that, power electronic interface units play an important role in maintaining the interaction with the main grid.

Operational requirements: When solar PV generation is available, maximum power point tracking (MPPT) is required to enhance the efficiency. MPPT is an important concept related to solar PV systems such that solar PV systems operate by extracting the maximum possible power at every instance. Most of the advanced power electronic interface units for solar PV systems have the MPPT feature.

4.1.2 DC TO DC CONVERTERS

A DC to DC converter is a power electronic interface that converts the voltage level of a DC source to another level. There are two main types of DC to DC converters, linear and switched, based on the method of conversion.

Linear DC to DC converters: As the name implies, a linear converter regulates the output DC voltage using a linear component. For example, a resistive load can be used in a linear DC to DC converter such that a resistive voltage drop regulates

DOI: 10.1201/9781003216292-4

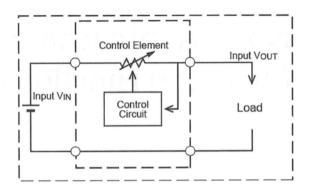

FIGURE 4.1   Linear DC to DC converter

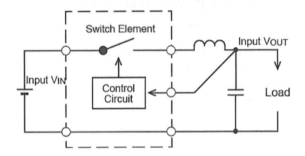

FIGURE 4.2   Switched DC to DC converter

the output voltage level. The general structure of a linear DC to DC converter is given in Figure 4.1.

Switched DC to DC converters: In switched mode converters, the input energy is stored in a periodical manner and released to obtain an output voltage at a desirable level. Inductors, transformers, and capacitors can be used as energy storing components. The general structure of a switched DC to DC converter is given in Figure 4.2.

To be specific, the input power supply is converted to a voltage pulse by the switching element in the circuit. The capacitors, inductors, and other elements smooth the generated voltage pulse. The power is supplied from the input side to the output by turning ON a switch (such as a MOSFET: a Metal–Oxide–Semiconductor Field-Effect Transistor, which is a field-effect transistor where the voltage determines the conductivity of the device, and which is used for switching or amplifying signals) until the desired voltage is reached. Once the output voltage reaches the predetermined value, the switch element is turned OFF and no input power is consumed. This operation is repeated at high speeds to improve the efficiency of the converter and to reduce thermal losses.

A comparison between linear and switched DC to DC converters is presented in Table 4.1.

TABLE 4.1

Comparison between Linear and Switched Converters

Linear Converters	Switched Converters
Simpler configuration	Complicated configuration
Requirement for few external components	Requirement for more external components
Lower noise	Higher noise
Only DC/DC step down (buck) operation	Supports buck, boost, buck–boost, etc. operations
Comparatively lower efficiency	Comparatively higher efficiency
Significant heat generation	High frequency switching operation lowers the heat generation

Switched DC to DC converters can be categorized based on the electrical isolation of the circuit, as shown in Figure 4.3. The term "isolation" refers to whether or not the input and the output of the DC to DC converter are electrically separated. In an isolated DC to DC converter, a transformer is included in the circuit such that the DC path between the input and the output is eliminated. Unlike an isolated converter, a non-isolated DC to DC converter consists of a DC path between the input and the output.

Let's analyze the characteristics and operation of buck, boost and buck–boost converters. In DC to DC converters, the operation of semiconductor switches such as insulated gate bipolar transistors (IGBTs) or MOSFETs occur periodically. Let's take the operational period as T_s. The switch is closed and opened for times T_{ON} and T_{OFF} respectively. The duty ratio D of the converter is the ratio between the active time of the converter and the switching period, as given in equation (4.1).

$$D = \frac{T_{ON}}{T_s} = \frac{T_{ON}}{T_{ON} + T_{OFF}}$$

(4.1)

4.1.2.1 Buck Converter

A buck converter reduces the input DC voltage to a specified DC output voltage. A typical buck converter is shown in Figure 4.4.

The input voltage source is coupled to a switch-like controlled solid-state device. A Power MOSFET or an IGBT can be used as the solid-state device. Thyristors are not commonly used in DC to DC converters because turning off a thyristor in a DC to DC circuit necessitates another commutation, which necessitates the use of another thyristor, whereas Power MOSFETs and IGBTs can be turned off by simply setting the voltage between the GATE and SOURCE terminals of the Power MOSFET, or the GATE and COLLECTOR terminals of the IGBT, to zero.

A diode is utilized as the second switch. A low-pass LC filter (LC filters are circuits consisting of a combination of inductors (L) and capacitors (C) to cut or

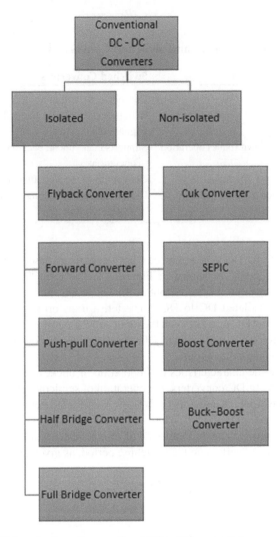

FIGURE 4.3   Categorization of conventional DC to DC converters

pass specific frequency bands of an electric signal) is used to connect the switch and the diode, and is designed to reduce current and voltage ripples. The load is completely resistive. The input voltage is constant, as is the current flowing through the load. The current source can be considered as the load.

Pulse width modulation (PWM) is used to turn on and off the controlled switch. PWM can be based on time or frequency. The use of a wide range of frequencies to accomplish the desired control of the switch, which in turn gives the desired output voltage, is a disadvantage of frequency-based modulation. This necessitates a sophisticated design for the low-pass LC filter, which must handle a wide frequency range.

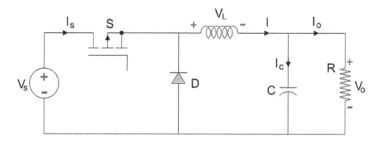

FIGURE 4.4 Buck DC/DC converter topology

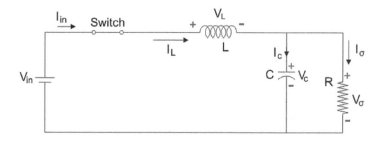

FIGURE 4.5 Switch–on state, diode–off state

DC to DC converters commonly use time-based modulation. This is easy to put together and use. In this sort of PWM modulation, the frequency remains constant.

The operation of a buck converter can be discussed for two operating modes as follows.

Mode 1: Switch–ON State, Diode–OFF State

The circuit configuration of a buck converter that operates in Mode 1 is shown in Figure 4.5. Let's derive a mathematical model for the buck converter operation in this mode of operation. Here the voltage across the capacitance can be considered to be equal to the output voltage under the steady state condition.

Here we can consider that the switch is on for a time T_{ON} and is off for a time T_{OFF}. We define the time period, T, as in equation (4.2)

$$T = T_{ON} + T_{OFF} \tag{4.2}$$

and the switching frequency is given in equation (4.3),

$$f_{switching} = \frac{1}{T} \tag{4.3}$$

The duty cycle D was previously defined by equation (4.1),

$$D = \frac{T_{ON}}{T_s} = \frac{T_{ON}}{T_{ON} + T_{OFF}}$$

For the buck converter operation in steady state for this mode, using Kirchhoff's Voltage Law,

$$\therefore V_{in} = V_L + V_o \tag{4.4}$$

$$\therefore V_L = L\frac{di_L}{dt} = V_{in} - V_0 \tag{4.5}$$

$$\frac{di_L}{dt} = \frac{\Delta i_L}{\Delta t} = \frac{\Delta i_L}{DT} = \frac{V_{in} - V_0}{L} \tag{4.6}$$

Considering the switch close time $T_{ON} = DT$ we can say that $\Delta t = DT$.

$$\left(\Delta i_L\right)_{closed} = \left(\frac{V_{in} - V_0}{L}\right)DT \tag{4.7}$$

Mode 2: Switch–OFF State, Diode–ON State
Here the energy stored in the inductor is released and dissipated in the load resistance, which aids in maintaining current flow through the load. However, when analyzing the circuit with Kirchoff's Voltage Law, we stick to the original conventions.

The circuit configuration of a buck converter that operates in Mode 2 is shown in Figure 4.6. For the buck converter operation in steady state for this mode, using Kirchhoff's Voltage Law,

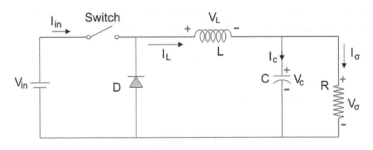

FIGURE 4.6   Switch–off state, diode–on state

$$0 = V_L + V_o \tag{4.8}$$

$$\therefore V_L = L\frac{di_L}{dt} = -V_o \tag{4.9}$$

$$\frac{di_L}{dt} = \frac{\Delta i_L}{\Delta t} = \frac{\Delta i_L}{(1-D)T} = \frac{-V_o}{L} \tag{4.10}$$

Considering the switch open time,

$$T_{OFF} = T - T_{ON} = T - DT = (1-D)T \tag{4.11}$$

Taking $\Delta t = (1-D)T,$

$$\left(\Delta i_L\right)_{open} = \left(\frac{-V_o}{L}\right)(1-D)T \tag{4.12}$$

The net change of the inductor current throughout any complete cycle is already known to be zero.

$$\left(\Delta i_L\right)_{closed} + \left(\Delta i_L\right)_{open} = 0 \tag{4.13}$$

$$\left(\frac{V_{in} - V_0}{L}\right)DT + \left(\frac{-V_o}{L}\right)(1-D)T = 0 \tag{4.14}$$

$$\frac{V_o}{V_{in}} = D$$

4.1.2.2 Boost Converter

A boost converter converts a specific DC input voltage to a DC output voltage. Figure 4.7 shows an example of a standard boost converter.

A mathematical model can be derived for the boost converter in a similar way to the previously discussed buck converter, such that the output voltage is given as,

$$\frac{V_o}{V_{in}} = \frac{1}{1-D}$$

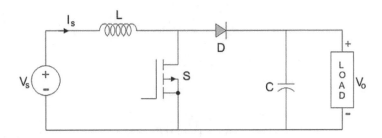

FIGURE 4.7   Boost DC/DC converter topology

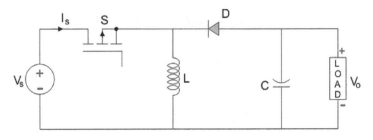

FIGURE 4.8   Buck–boost DC/DC converter topology

4.1.2.3   Buck–Boost Converter

A buck–boost converter has an output voltage magnitude that is either greater than or less than the input voltage magnitude.

The circuit configuration of a buck–boost converter is shown in Figure 4.8. A mathematical model can be derived for the buck–boost converter in a similar way to the previously discussed buck converter, such that the output voltage is given as,

$$\frac{V_o}{V_{in}} = \frac{D}{1-D}$$

Activity: Derive the mathematical models for the boost and buck–boost converters following the buck converter modeling.

4.1.3   DC to AC Inverters

4.1.3.1   Voltage Source Inverter

A VSI is composed of a DC voltage source, a three-phase bridge circuit, and a capacitor with a significantly greater capacitance, as shown in Figure 4.9.

As listed below, there are two varieties of VSI available.

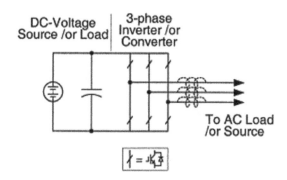

FIGURE 4.9 Structure of voltage source inverter

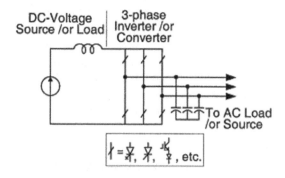

FIGURE 4.10 Structure of current source inverter

4.1.3.1.1 Square-Wave Inverter Using Gate Turn-Off Thyristor

The DC capacitor input voltage determines the AC output voltage of this converter, hence the AC output voltage can be varied by adjusting the DC input voltage. The proportionality between the input voltage and the basic component of the output voltage is the main principle of a VSI.

4.1.3.1.2 PWM Inverter Using Insulated Gate Bipolar Transistor

Unlike the previous VSI, this one has a constant DC input voltage. The output is modified using the PWM approach, in which the modulator's modulation index is changed according to the required AC output voltage.

4.1.3.2 Current Source Inverter

As shown in Figure 4.10, a CSI consists of a DC source, a three-phase bridge, and a DC source that feeds the main converter circuit. A somewhat large DC inductor fed by a voltage source can be used as the DC source.

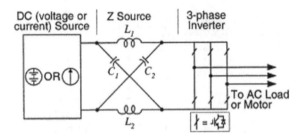

FIGURE 4.11   Structure of Z source inverter

4.1.3.3   Z Source Inverter

A ZSI is composed of a DC source, a two-port network made up of a split-inductor (two inductors, L1, L2), and two capacitors (C1, C2) coupled in an X shape as in Figure 4.11, and it feeds the main converter circuit with DC via impedance source coupling. A voltage or current source might be used as the DC source. A ZSI provides a simplified single-stage power conversion topology as an added benefit. The shoot-through (ST) can no longer kill the inverter, and it adds DC to DC power conversion (buck–boost mode) to the inverter in addition to DC to AC power conversion. It solves the majority of the issues of traditional voltage and current source inverters.

4.1.4   RECONFIGURABLE POWER AND CONTROL ARCHITECTURES OF MICROGRIDS

4.1.4.1   Reconfigurable Systems

Reconfigurable systems can change their functional capability by attaining alternative configurations at different times. Such systems are especially suited to certain types of applications, where their ability to adapt easily to changing conditions can be used to meet new needs and improve survivability by increasing reliability. Microgrid architectures are primarily divided into two categories: control architecture and power architecture.

4.1.4.2   Existing Power Architecture-Based Reconfigurable Approaches for Microgrids

To build and create innovative reconfigurable power architectures for solar PV systems and microgrids, numerous studies have been conducted. Reconfigurable solar converter (RSC) based power conversion architectures have been presented, with the goal of adding technical, financial, and economic value to existing solar PV systems [1–3]. This design is still being developed in order to create new power architectures for solar farms and AC/DC hybrid residences [4, 5].

The nature of microgrids is being able to switch between grid-connected and islanded modes by altering the operating mode, and there are numerous studies on this topic [6]. Distribution feeder reconfiguration to form multiple microgrids is proposed as the main consideration in optimum microgrid planning under uncertainties [7], by taking into account economic, technical, and reliability aspects under uncertainties in load demand, electricity price, and renewable power generation.

4.1.4.3 Existing Control Architecture-Based Reconfigurable Approaches for Microgrids

A new reconfigurable architecture for a microgrid powered by photovoltaic, wind, micro hydro, and fuel cell-based power and with power backup from ultra-capacitor (UC) and battery storage has been proposed [8], taking into account existing centralized, decentralized, and hierarchical microgrid control architectures. Through reconfigurability, the main goal of this novel architecture is to boost microgrid dependability by building a microgrid with high fault tolerance to failure or malfunctioning of microgrid controllers and communication links.

The solar inverter is the most controllable component of a solar PV system, and it will be a fundamental component of the reconfigurable design. As a result, solar inverters should be designed to function in a certain operating mode and transition to the most efficient mode of operation automatically.

4.1.5 Modeling of Solar Microgrids with a Z Source Inverter

The general structure of a solar microgrid with a Z source inverter is given in Figure 4.12.

High solar PV penetration, according to the grid, generates more obstacles, such as power quality issues, intermittent power generation, and issues with reliable power delivery upon system failures. As a result, voltage regulation and

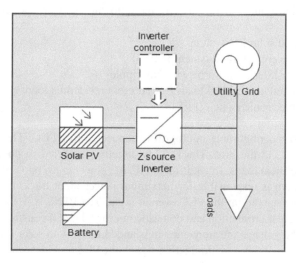

FIGURE 4.12 General structure of a solar microgrid with a Z source inverter

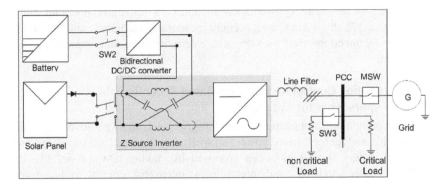

FIGURE 4.13   Proposed reconfigurable architecture for residential microgrid

harmonic mitigation have become essential for solar PV systems. In addition to power generation, the reactive power regulation function of solar inverters at night is being researched, and has been effectively confirmed, in both technical and economic terms, by studies.

ZSIs are the current focus of solar PV inverter research because they reduce power losses by reducing the power conversion stage, improve power quality by minimizing output voltage and current harmonics by reducing dead time, and improve inverter reliability by reducing inverter failures due to unavoidable shoot-through conditions and capacitor failures due to higher voltage stress.

4.1.5.1   Example of Proposed System with a ZSI

The main aim of this study is to rearrange the power architecture of a solar PV microgrid. Figure 4.13 depicts the proposed reconfigurable architecture of a residential microgrid. It is composed of the following main components:

- ZSI-based solar PV system
- Battery energy storage system (BESS)
- LC ripple filter (to absorb switching ripple)
- Critical and non-critical loads (to represent residential loads)
- Power electronic switch (PES)

The proposed reconfigurable microgrid is connected to a PCC. The PES is used to link the PCC to the grid. The PCC is directly connected to both the critical and the non-critical loads. In addition to DC/AC power conversion, the ZSI-based solar PV system is responsible for maximum power point tracking (MPPT) and ensuring power quality at the PCC even at night. The BESS is used to bridge the gap between solar production and domestic microgrid (MG) consumption, as well as to adjust the voltage and frequency in islanded mode with solar PV. Depending on the condition of the synchronization signal, the PES provides smooth switching (ON/OFF) of the reconfigurable system.

4.1.5.2 Modes of Control of a ZSI

A summary of the modes of control of a ZSI can be illustrated as in Figure 4.14.

4.1.5.2.1 When the Solar Microgrid Is Operated in Grid-Connected Mode

A residential MG is normally operated as a grid-connected microgrid, with the inverter controller set to current controlling mode as in Figure 4.15. The ZSI is used to control the solar PV system as a current source, which feeds the microgrid. The major control goal of this mode is to extract as much power as possible from the solar PV. The batteries and the utility grid meet the power demand of the residential loads when there is a power deficit. If the solar PV generates too much energy, the BESS is used to store it while the rest is sent to the grid. The configuration and power flow in the "Grid-connected microgrid" mode are given in Figure 4.15.

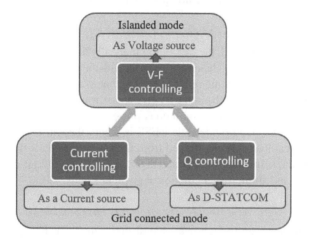

FIGURE 4.14   States of control of solar PV system

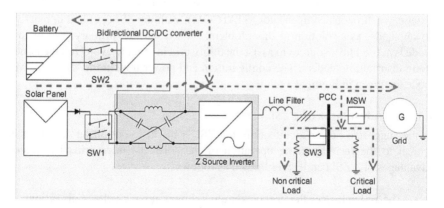

FIGURE 4.15   Configuration and power flow for the "Grid-connected microgrid" mode

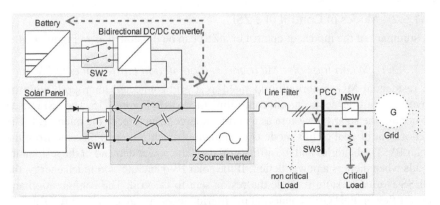

FIGURE 4.16 Configuration and power flow for the "Islanded microgrid" mode

4.1.5.2.2   When the Solar Microgrid Is Islanded

When a grid fault is detected, the MG is automatically disconnected from the utility grid and reconstructed as an islanded microgrid, with the inverter controller switched to voltage-frequency regulating mode, ensuring that local loads are not interrupted. The ZSI is used to operate the solar PV system as a voltage source. The primary goal of this mode is to deliver continuous power to key loads while maintaining power quality in islanded mode. The BESS is where the extra solar electricity is kept. Non-critical loads are unplugged from the MG if the MG's demand exceeds the power available from the battery and the solar PV. The configuration and power flow in the "Islanded microgrid" mode are given in Figure 4.16.

4.1.5.2.3   At Night When Solar Generation Is Unavailable

Because solar generation is not available at night, the solar PV system is turned off. However, due to the Ferranti effect, the voltage rises at night on most lengthy distribution feeders in rural distribution networks. As a result, improper microgrid tripping may occur at any time. The solar PV microgrid is redesigned to operate as a static synchronous compensator (STATCOM), with the ZSI operated in reactive power mode, as a solution to this problem. As a result, the primary goal of this mode is to use idle resources to assist the utility grid in order to avoid voltage issues along distribution feeders. The configuration and power flow in the "STATCOM" mode are given in Figure 4.17.

4.1.5.3   Advantages of a ZSI

In the suggested power architecture, the ZSI is the key controlling device. It is critical to the reconfigurable operation of the solar PV microgrid. The major advantages of a ZSI can be highlighted as follows.

- The ZSI enhances the grid power quality by removing voltage distortions
- When solar power is available, the maximum amount of power can be generated by the solar PV system

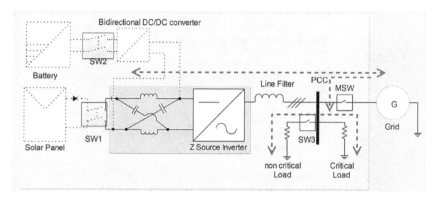

FIGURE 4.17 Configuration and power flow for the "STATCOM" mode

- The ZSI's ability to provide reactive power boosts grid stability and provides sustained value to system owners during non-production hours
- The reactive power control mode is used to improve the voltage profile at night and save money on expensive capacity banks

4.2 ADAPTIVE PROTECTION FOR MICROGRIDS

4.2.1 OVERVIEW OF POWER SYSTEM PROTECTION

The detection and disconnection of short circuits and other abnormal circumstances in the electrical supply is known as protective relaying or protection. Power system protection is an important consideration in electrical engineering for preventing or reducing harm to a power system's critical elements. Furthermore, a correctly designed protection plan is critical for assuring system reliability, protecting human safety, minimizing power disruptions, and maintaining supply quality, among other things.

4.2.1.1 Protection System Components

As indicated in Figure 4.18, a traditional protective system consists primarily of sensing, decision-making, switching, power supply, communication, and control elements. Sensing devices are primarily designed to deliver accurate feedback on the power system's desired operation, via continuous monitoring. The electric signals produced by sensing devices activate decision-making devices such as relays. When a faulty state is recognized, relays send out protection signals, which open or close the faulty circuit as needed. Circuit breakers are in charge of breaking circuits carrying fault currents, based on feedback from relays. Batteries and other power supply devices are used to ensure that the relays and circuit breakers in the system have an uninterrupted power supply and are not reliant on the system's primary power source.

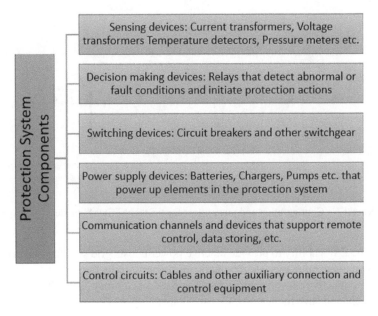

FIGURE 4.18 Protection system components

4.2.1.2 Properties of a Protection System

Sensitivity, selectivity, reliability, and speed are the four essential characteristics of a protective system. All of these properties should be present in an ideal protection solution.

Sensitivity: This refers to the ability to detect the smallest fault conditions as well as moderate and severe defects. With large variations in the short circuit current level, sensitivity can be a serious issue in microgrid scenarios. The sensitivity of a relay's pickup current setting is directly related to the sensitivity of the protective system, and it should be adjusted properly to provide protection at all fault levels.

Selectivity: This refers to how well primary and backup protection mechanisms work together. In the event of a fault, the primary relays should be activated first, followed by backup protection with the required delay.

Reliability: Dependability and security are the two major sub-qualities of reliability. Dependability is the ability to perform correctly for the specified period of time. The capacity to avoid unnecessary activities under normal operating conditions is referred to as security. The protection, for example, should not be triggered by faults outside its zone and should have a transient tolerance.

Speed: Another important component in improving the performance of a protective scheme is speed. Faster problem resolution can help to keep network and equipment damage to a minimum. Unnecessarily speedy procedures can result in undesirable outcomes; therefore, tolerance levels should be established.

4.2.2 PRESENT MICROGRID PROTECTION SCHEMES

Microgrids are typically looped or meshed systems with distributed generation. They can operate in a variety of topologies, including grid-connected and islanded modes. These potential network alterations could result in variations in the short circuit level, necessitating constant monitoring of protection settings.

4.2.2.1 Line Protection

The most popular types of line protection are overcurrent and distance. The overcurrent method is the most straightforward and cost-effective of the two. Even though it is the simplest approach, protecting novel distribution networks with varying topologies and embedded generation is more complex.

Because of the meshed topology of these networks and the presence of distributed energy resources, directional overcurrent relays (DOCRs) are being used more frequently. The purpose of DOCRs is to determine the direction of current flow by measuring the phasor difference between current and voltage. The relay will only turn on if the fault is occurring in front of it. In a mesh system, this directionality function is critical for correctly locating the causes of failures. Relays will not operate if fault currents are flowing in the opposite direction.

4.2.2.2 Primary and Backup Protection

The microgrid primary relay is the one that is closest to the fault. A fault can have multiple primary relays, each of which isolates the fault by disconnecting a particular segment of the network. If the fault is not removed by the primary protection, backup protection steps in. Backup relays are normally time-synchronized with primary relays, and if the fault lasts longer than the predetermined duration, backup relays are triggered. Relay miscoordination can lead backup relays operating before primary relays, resulting in needless disconnections and the de-energization of large sections of the network. In a radial network, primary backup coordination is generally simple, but it can be time-consuming in a meshed or DG-sourced network.

4.2.3 ADAPTIVE PROTECTION SCHEMES FOR MICROGRIDS

Microgrids are dependable and cost-effective platforms for distributed generation (DG). They can work in both grid-connected and islanded configurations. As a result, there may be concerns with microgrid protection systems, such as fluctuations in fault current levels during the two modes of operation. This problem can be solved using an adaptive protection mechanism. The relay settings will be adjusted according to the operating mode under this scheme. Intelligent relays that can identify the mode of operation can be used for this purpose.

The fault current is the most significant distinction between these two modes [9]. Because the main utility grid produces a large quantity of fault current, the grid-connected mode has a high fault current level. The utility grid is separated from the microgrid during the islanded mode of operation, and it no longer serves as a fault current resource. As a result, there will be a change up to a particular level in the short circuit level. As a result of these considerations, the protective system should be adjusted to the mode of operation [10].

4.2.3.1   What Is Adaptive Protection?

Adaptive protection can be described as an online system that injects the externally generated signals to the power grid according to the changes in the system [11]. Microgrid circumstances are constantly altering in response to the fault level of the distribution network [12].

To handle these dynamic conditions in the power system, a novel adaptive relay model has been designed. Microprocessor-based relays offer multi-functional capabilities and are well-suited to such dynamic scenarios. They can continuously monitor the state of the electrical system and detect any disturbance within their protection range. Relays with adaptive features can modify their online status based on the microgrid's failure level.

4.2.3.2   Adaptive Protection Algorithms

This section introduces general algorithms for designing an adaptive protection scheme, including the relay operation and adaptive protection function in grid-connected and islanded modes. Figure 4.19 represents the block diagram for the adaptive protection relay used here. The operation algorithms for islanded and grid-connected modes are shown in Figures 4.20 and 4.21 respectively.

The protective elements that are necessary for this scheme are comparable to those required for a traditional grid, such as transducers, circuit breakers, and relays. As the microgrid's operating mode shifts from grid-connected to islanded,

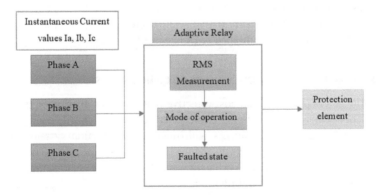

FIGURE 4.19   Block diagram of adaptive protection relay

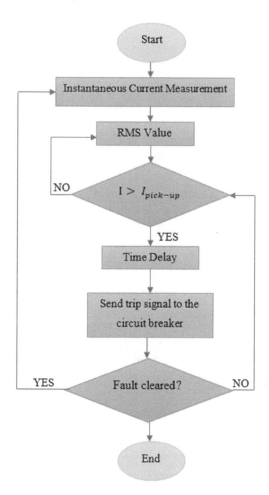

FIGURE 4.20   Flow chart for islanded mode

the fault level drops dramatically. As a result, it is critical that the relay understands the microgrid's operating mode.

4.2.4 CASE STUDY

Here we consider a research-based outcome of the implementation of an adaptive protection scheme for a microgrid. The test system in Figure 4.22 is applied in Sri Lanka, taking into consideration a real scenario. The electric grid operates at 400 V and is stepped down to 230 V. The distributed generating units that make up the microgrid generate 230 V electricity. Through the PCC, the microgrid is connected to the main utility. The PCC is controlled by a breaker, which opens when the islanded mode of operation occurs.

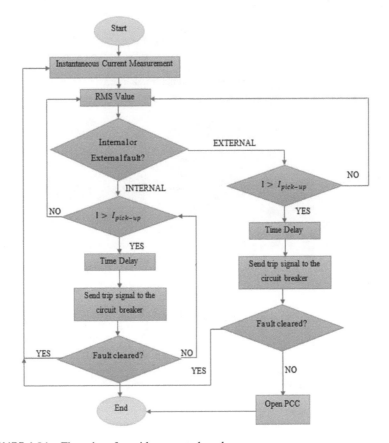

FIGURE 4.21   Flow chart for grid-connected mode

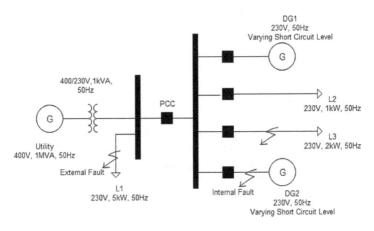

FIGURE 4.22   Microgrid test system

The system operation can be tested under two scenarios as follows.

Case 1: Fault at distributed generator
Case 2: Fault at the load on the microgrid side

When the PCC is in the islanded mode, it is open; when it is in the grid-connected mode, it is closed. In MATLAB, we have a circuit breaker as a PCC, which is manually controlled. The problem occurs near DG1 on the microgrid side at point A, as illustrated in Figure 4.22. On the microgrid side, another problem occurs at point B, which is close to load 3 (L3). During these two tests, measurements of the power system current and fault current are taken in a variety of fault scenarios:

- Three-phase to ground fault
- Phase to phase fault

TABLE 4.2
Test Results

Category	Maximum Load Current (A)	Maximum Fault Current (A)	Fault Clearing Time (s)
Three-phase to ground fault			
Near DG in islanded mode	5	27.664	0.428
Near DG in grid-connected mode	12	60	0.350
Near the load on microgrid side in islanded mode	5	27	0.435
Near the load on microgrid side in grid-connected mode	10	47	0.360
Two-phase to ground fault for the two test cases			
Near DG in islanded mode	5	28	0.435
Near DG in grid-connected mode	10	60	0.336
Near the load on microgrid side in islanded mode	6.5	55	0.362
Near the load on microgrid side in grid-connected mode	5.7	85.5	0.326
Single-phase to ground fault for the two test cases			
Near DG in islanded mode	5	27	0.435
Near DG in grid-connected mode	10	47	0.360
Near the load on microgrid side in islanded mode	6.5	55	0.355
Near the load on microgrid side in grid-connected mode	5.8	72.5	0.335
Phase to phase fault for the two test cases			
Near DG in islanded mode	5	25	0.433
Near DG in grid-connected mode	10	60	0.355
Near the load on microgrid side in islanded mode	6.5	49	0.365
Near the load on microgrid side in grid-connected mode	6	79	0.334

- Single-phase to ground fault
- Two-phase to ground fault

These data can be used to assess the effectiveness of the adaptive protection scheme that has been built. Table 4.2 displays the results obtained.

According to the results obtained, we can observe that:

- Depending on the mode of operation, the fault current level varies significantly
- The time it takes for the relay to trip is reasonable in relation to the system's fault level
- As the fault current increases, the fault clearing time decreases

The major goal of this study is to create a novel adaptive microgrid protection method based on the overcurrent principle and the level of fault current in the microgrid, by adjusting the relay settings based on the mode of operation. The results of the tests indicate that the proposed relay system operates appropriately and provides adequate protection in both modes of operation. Furthermore, if the relay's time delay could be decreased, the relay's performance would be improved.

4.3 MULTI-AGENT-BASED CONTROL

4.3.1 INTRODUCTION TO MULTI-AGENT SYSTEMS

Basically, an agent is a software entity or a computer system that is placed in a particular environment and is capable of performing autonomous actions within the environment to fulfil assigned tasks or duties [13, 14]. An agent is capable of taking sensory inputs from its environment and outputs actuating signals as in Figure 4.23.

Generally, an agent can be represented by a controller. Agents have inherent qualities such as *autonomy*, which means functioning without human intervention, *social ability*, which is the ability to cooperate with humans and

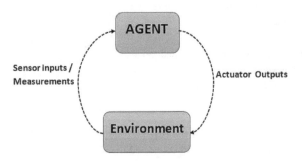

FIGURE 4.23 An agent in its environment

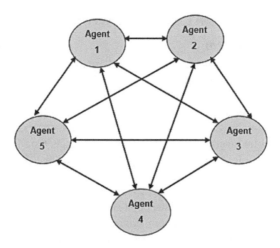

FIGURE 4.24   Agents in a MAS

other agents, *reactiveness*, which allows them to respond quickly to changes in their environment, and *proactive behavior*, which allows them to achieve their goals [15]. An agent can have extra features such as *mobility*, which is the ability of the agent to travel between nodes of a computer network, *benevolence*, which denotes the agent's ability to certainly perform what it is assigned to do, *rationality*, meaning the agent will achieve goals without preventing the achievement of common tasks, and *truthfulness*, ensuring that the false information will not be provided. Interaction between agents in a multi-agent system (MAS) is shown in Figure 4.24.

A MAS consists of multiple agents carrying out tasks. Agents in a multi-agent system interact with each other in solving problems, since the particular task can be beyond the capability of any individual agent. Agents in a system can be different from each other or they can be identical in their tasks, available data, actions, measurements etc., so that there are two types of typical agents in a system – heterogeneous (different) and homogeneous (identical) agents. Depending on the inputs that they receive, the agents communicate with each other. This communication can take place by transmitting information directly between each agent in the system or through a central directory. It is also not essential for each and every agent to communicate with each other agent, and there can be non-communicative agents.

As opposed to a centralized approach, the decentralized design of a MAS does not suffer from single point failure. In terms of computational efficiency, dependability, extensibility, resilience, maintainability, flexibility, and reuse, the introduction of a MAS into traditional applications has improved total system performance. Aircraft maintenance, electronic book purchasing, supply-chain management, microgrid energy management techniques, and other MAS research areas might be included [16, 17].

4.3.2   MULTI-AGENT-BASED CONTROL FOR MICROGRIDS

4.3.2.1   Proposed System

The proposed microgrid structure is a single-phase system. A solar PV unit, wind sources and a battery backup are connected to the system as distributed energy resources. Two load sets are connected separately with critical loads and non-critical loads.

During the grid-connected operation, DC sources are connected to the AC bus bar through a grid-tie inverter. When there is a grid outage, the MAS maintains the supply to the critical loads through a standalone inverter which is connected to a solar panel through a battery backup and a charge controller.

We are concerned with a grid-connected microgrid. The main aim of this microgrid operation is to limit the power consumption from the grid by introducing renewable energy sources to the system, considering the current demand. Here we consider two different energy sources, such as wind turbines and solar photovoltaic sources. Then, by defining a priority order for the switching of these sources with increasing demand we can achieve energy management functions. In addition, when the supply from the renewable energy sources is low, non-critical loads in the system are to be shed accordingly. It is proposed that if the grid fails, the microgrid will be isolated by tripping the connected solar panels and wind turbines and switching to a diesel generator.

Let's consider a microgrid for a manufacturing industry. Here the main aim is to limit the power consumption from the grid by adding renewable sources accordingly. Let's assume that there are two separate loads to be supplied and that two main renewable sources (a wind turbine and solar photovoltaic panels) are available. Considering the unit cost of each renewable energy source, depending on the variety of their technologies, a priority order can be specified.

Let's suppose that the above unit costs are specified for Sri Lanka, and then a priority order can be specified depending on the unit cost. The wind turbine is to be connected initially to satisfy the load demand, and then as the demand increases solar sources will be added to the system. On a typical day it is proposed to maintain the loads within the limit of the power available from renewable sources. The two load sets can be identified as critical loads and non-critical loads. The critical loads are the loads that directly affect the ability of the manufacturing firm to maintain its key operations. As the name implies, the non-critical or non-essential loads are the loads that can be cut off if there is insufficient power, such as office lighting, printers, fans etc. Therefore, the non-critical load set is to be shed, giving priority to the critical loads. The non-critical loads should not be powered up until a certain low demand is achieved.

4.3.2.2   Agents in the System and Their Functions

As described in the previous section, there are four main types of agent available in the proposed system: solar agent, wind agent, load agent and server agent. The functions of each type of agent can be stated as follows.

- **Wind Agent**

  Current and voltage measurements are taken when connected to the bus bars. The wind agent receives ON/OFF signals depending on the system requirement and its availability. When the agent is switched on or off, the corresponding ON/OFF status signals will be sent to the server agent. The wind agent has a display unit. Current and voltage measurements will be displayed through this unit.

- **Solar Agent**

  Current and voltage measurements are taken when connected to the system. The solar agent receives ON/OFF signals depending on the system requirement and its availability. When the agent is switched on or off, the corresponding ON/OFF status signals will be sent to the server agent. The solar agent has a display unit. Current and voltage measurements will be displayed through this unit. When solar power is available, that particular status signal will be sent to the load agents allowing them to make their demands.

- **Load Agent**

  Current and voltage measurements are taken when connected to the bus bars. When loads are connected or disconnected, the corresponding status signals will be sent to the server agent. Each load agent has a display unit which displays current and voltage measurements.

- **Server Agent**

  This agent follows the main controlling algorithm and accordingly sends ON/OFF signals to the solar panels. Also, when loads are connected or disconnected, the corresponding status signals and the total power requirement will be received by the server agent.

In the proposed system there are two load agents, a wind agent, a solar agent and a server agent. The algorithms for the system operation are described in the following sections.

4.3.3 Simulating the Interaction between Agents Using JAVA Agent Development Environment

4.3.3.1 JAVA Agent Development Environment

Java Agent Development Environment (JADE) is a software platform which is implemented in Java. This provides a user-friendly platform for the creation and execution of the agents in a MAS, since interactions between agents in a MAS take the form of a mesh which is completely through communication channels. In order to demonstrate these interactions, graphical toolkits are provided by JADE. This software platform also supports the standard Agent Communication Language (ACL) implementation. The basic configuration of JADE platforms is as in Figure 4.25.

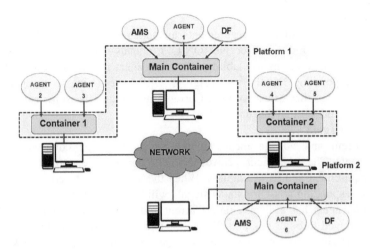

FIGURE 4.25   JADE architecture

The JADE platform is composed of multiple agent "containers". These containers are generally distributed over the networks. Agents are created and exist inside these containers. A container can be defined as a Java process that assists in all the services related to the hosting and execution of agents. Among these multiple containers, there is a special container called the "main container". Generally, this is the first container to be formed within the system; other containers will then gradually be added to the system by being registered under the main container. A programmer can identify each and every container in the JADE platform according to a specified logical name. As an example, the initial container is named the "Main container", and then the additional containers are named "Container 1", "Container 2", etc. The main container has special objectives such as maintaining the agent management system (AMS) and the directory facilitator (DF). The AMS and the DF are two special agents that are responsible for the white page service and the yellow page service, respectively. In addition to that, the main agent is responsible for managing the global agent descriptor table and the container table. The global agent descriptor table is a registry that contains the current status and location of each agent in the system. The container table is a registry which is used to store object references of all nodes in the JADE platform.

4.3.3.2   Agent Formation

As the Java coding platform, IntelliJ IDEA COMMUNITY Version 2019.1.3 is used. The start-up screen is shown in Figure 4.26.

"Agent Class" allows the creation of JADE agents. The class provides basic methods for performing the actions of the agents, such as passing messages as ACL messages, supporting the life cycle of an agent, planning and executing multiple activities simultaneously, and so on.

FIGURE 4.26 Java coding platform IntelliJ IDEA COMMUNITY Version 2019.1.3

In our simulation there are two main classes, the power source class and the consumer class, which are used to create, respectively, source agents and load agents in the system, as given in Figure 4.27. The assigned tasks for each agent are specified through "Agent Behavior", which defines the actions under a given event.

4.3.3.3   Sniffing Agent

As mentioned in the above sections, the JADE platform provides a graphical toolkit to demonstrate agent communication. Most of the available tools are used in debugging a particular agent. The "sniffing tool" is dedicated to debugging and keeping a track of conversations between agents. This track of information kept by the sniffing agent includes messages exchanged between the dedicated agents in the system and all platform events. The sniffing agent is also created by the "Agent Class" tool which is provided through the graphical user interface of JADE as in Figure 4.28.

The left pane of the graphical user interface shows the agent platform, allowing the user to select agents for sniffing. The right pane graphically represents the sniffed agents in blocks. Each arrow shows a message, and arrows of the same shade denote a particular conversation. When the user selects the agents to be sniffed, each and every message forwarded to or coming from those agents is displayed on the graphical user interface. This allows the user to have a detailed description of each and every message that has taken place, and the messages are saved in a text file format or in binary format for later retrieval.

For our proposed system, the sniffing tool shows the available source agents as the wind and solar agents. Load agents are also added to the sniffing platform. When the available power of the generating sources and the connected loads are

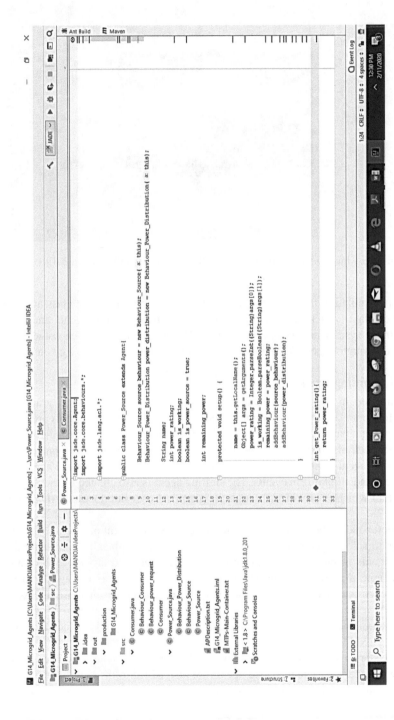

FIGURE 4.27 Power and consumer source classes formation in JADE interface

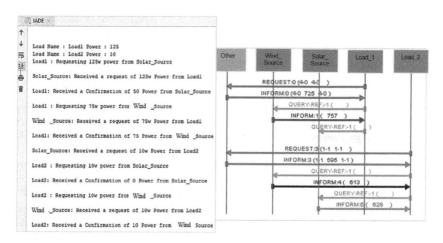

FIGURE 4.28   Simulation through the sniffing tool

specified, the agents will communicate with each other to supply the requested demand. Proposals, requests and queries which are exchanged among agents in the system are graphically shown through the sniffer graphical user interface.

REFERENCES

[1] H. Kim, B. Parkhideh, T. D. Bongers, and H. Gao, "Reconfigurable solar converter: A single-stage power conversion PV-battery system," *IEEE Trans. Power Electronics*, vol. 28, no. 8, pp. 3788–3797, 2013, doi: 10.1109/TPEL.2012.2229393

[2] I. Mazhari, M. Chamana, B. H. Chowdhury, and B. Parkhideh, "Distributed PV-battery architectures with reconfigurable power conversion units," *IEEE Appl. Power Electron. Conf. Exposition*, pp. 691–698, 2014, doi: 10.1109/APEC.2014. 6803383.

[3] N. Sasidharan and J. G. Singh, "A novel single-stage single-phase reconfigurable inverter topology for a solar powered hybrid AC/DC home," *IEEE Trans. Ind. Electron.*, vol. 64, no. 4, pp. 2820–2828, 2017, doi: 10.1109/TIE.2016.2643602

[4] R. Rizzo, I. Spina, and P. Tricoli, "A single input dual buck–boost output reconfigurable converter for distributed generation," *5th Int. Conf. Clean Elec. Power, ICCEP* 2015, pp. 767–774, 2015, doi: 10.1109/ICCEP.2015.7177579

[5] A. Chub, D. Vinnikov, R. Kosenko, and E. Liivik, "Wide input voltage range photovoltaic microconverter with reconfigurable buck–boost switching stage," *IEEE Trans. Ind. Electron.*, vol. 64, no. 7, pp. 5974–5983, 2017, doi: 10.1109/ TIE.2016.2645891

[6] K. M. Reddy and B. Singh, "Multi-objective control algorithm for small hydro and SPV generation based dual mode reconfigurable system," *IEEE Trans. Smart Grid*, vol. 9, no. 5, pp. 4942–4952, 2018, doi: 10.1109/TSG.2017.2676013

[7] E. Ghiani, S. Mocci, and F. Pilo, "Optimal reconfiguration of distribution networks according to the microgrid paradigm," *2005 IEEE Int. Conf. Future Power Systems*, 2005, doi: 10.1109/FPS.2005.204290.

 [8] S. S. Thale, R. G. Wandhare, and V. Agarwal, "A novel reconfigurable microgrid architecture with renewable energy sources and storage," *IEEE Trans. Ind. Appl.*, vol. 51, no. 2, pp. 1805–1816, 2015, doi: 10.1109/TIA.2014.2350083

 [9] K. Tazi, F. M. Abbou, and F. Abdi, "Multi-agent system for microgrids: design, optimization and performance," *Artificial Intelligence Rev.*, vol. 53, no. 2, pp. 1233–1292, 2020, doi: 10.1007/s10462-019-09695-7

[10] M. Amaratunge, D. U. Y. Edirisuriya, A. L. A. P. L. Ambegoda, M. A. C. Costa, W. L. T. Peiris, and K. T. M. U. Hemapala, "Development of adaptive overcurrent relaying scheme for IIDG microgrids," *2nd Int. Conf. Elec. Eng. (EECon)*, 2018, pp. 71–75, doi: 10.1109/EECon.2018.8540999

[11] W. L. T. Peiris, W. H. Eranga, K. T. M. U. Hemapala, and W. D. Prasad, "An adaptive protection scheme for small scale microgrids based on fault current level," *2nd Int. Conf. Elec. Eng. (EECon)*, 2018, pp. 64–70, doi: 10.1109/EECon.2018.8540992

[12] T. S. S. Senarathna, and K. T. M. U. Hemapala, "Review of adaptive protection methods for microgrids," *AIMS Energy*, vol. 7, no. 5, pp. 557–578, 2019, doi: 10.3934/energy.2019.5.557

[13] H. V. V. Priyadarshana, M. A. K. Sandaru, K. T. M. U. Hemapala, and W. D. A. S. Wijayapala, "A review on multi-agent system based energy management systems for micro grids," *AIMS Energy*, vol. 7, no. 6, pp. 924–943, 2019, doi: 10.3934/ENERGY.2019.6.924.

[14] M. K. Perera *et al.*, "Multi-agent based energy management system for microgrids," *PIICON 2020 – 9th IEEE Power India Int. Conf.*, pp. 1–5, 2020, doi: 10.1109/PIICON49524.2020.9113021.

[15] V. N. Coelho, M. Weiss Cohen, I. M. Coelho, N. Liu, and F. G. Guimarães, "Multi-agent systems applied for energy systems integration: State-of-the-art applications and trends in microgrids," *Appl. Energy*, vol. 187, pp. 820–832, 2017, doi: 10.1016/j.apenergy.2016.10.056.

[16] A. Dorri, S. S. Kanhere, and R. Jurdak, "Multi-agent systems: A survey," *IEEE Access*, vol. 6, pp. 28573–28593, 2018, doi: 10.1109/ACCESS.2018.2831228.

[17] A. González-Briones, F. De La Prieta, M. S. Mohamad, S. Omatu, and J. M. Corc hado, "Multi-agent systems applications in energy optimization problems: A state-of-the-art review," *Energies*, vol. 11, no. 8, pp. 1928, 2018, doi: 10.3390/en11081928.

5 Cyber Security for Smart Microgrids

5.1 OVERVIEW OF CYBER ATTACKS

A cyber attack is a harmful and purposeful attempt by an individual or organization to gain access to another person's or organization's information system. Typically, the attacker is aiming for financial gain from disrupting the victim's network [1–6].

5.1.1 TYPES OF CYBER ATTACK

5.1.1.1 Malware

The most common sort of cyber attack is a malware attack. Malware is harmful software, such as spyware, ransomware, a virus, or a worm, that is placed on a computer when a user clicks on a malicious link or email. Malware can block access to essential network components, harm the system, and collect confidential information, among other things, once it has gained access to the system.

5.1.1.2 Phishing

Cyber criminals send scam emails that appear to be from reputable sources. The victim is then misled into clicking on the malicious link in the email, which results in the installation of malware or the revealing of sensitive information such as credit card numbers and login credentials.

5.1.1.3 Man in the Middle Attack

When malicious hackers insert themselves in the middle of a two-way conversation, this is known as a Man in the Middle (MitM) attack. Once the attacker has decoded the message, they may filter and take sensitive information, as well as providing the user with various responses.

5.1.1.4 Denial of Service Attack

A Denial of Service (DoS) attack tries to overload systems, networks, or servers with traffic, rendering them unable to respond to valid requests. Multiple infected devices can also be used to launch an assault on the target system. A distributed denial of service (DDoS) attack is the term for this type of attack.

5.1.1.5 Ransomware

Ransomware is a sort of computer attack in which the attacker encrypts or locks the data of the victim and threatens to publish or prevent access to it unless a ransom is paid.

5.1.2 COMMON SOURCES OF CYBER THREATS

- Nation States
  Cyber attacks by a nation can have a negative impact by interrupting communications, military operations, and everyday life.
- Criminal Groups
  Criminal organizations seek to obtain financial advantage through infiltrating systems or networks. Identity theft, online fraud, and system extortion are all carried out by these groups via phishing, spam, spyware, and malware.
- Hackers
  Hackers use a variety of cyber methods to break through defenses and exploit flaws in computer systems and networks. Personal gain, revenge, stalking, financial gain, and political involvement drive them. For the thrill of a challenge or bragging rights in the hacking community, hackers create new forms of danger.
- Terrorist Groups
  Terrorists utilize cyber attacks to harm national security, compromise military equipment, disrupt the economy, and cause mass losses by destroying, accessing, or exploiting vital infrastructure.
- Hacktivists
  Hacktivists carry out cyber attacks in favor of political causes rather than for financial gain. They go against industries, organizations, and individuals who don't share their political beliefs.
- Malicious Insiders
  Employees, third-party vendors, contractors, or other business associates who have legitimate access to an enterprise's assets but misuse them to steal or destroy information for financial or personal benefit are considered insiders.

5.2 POWER ROUTING CONCEPT

The concept of power routing has added new features to power distribution systems in recent years. A power routing device not only monitors the current condition of the distribution network, but also collaborates with other devices to find alternative power flow channels if necessary, thanks to communication with generators, power lines, and consumers.

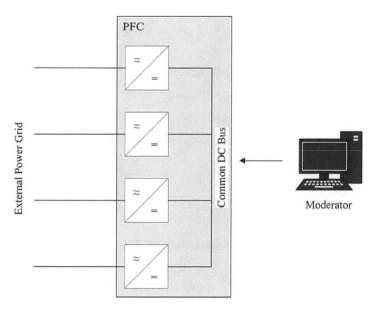

FIGURE 5.1   Power flow controller

The majority of power routing algorithms minimize load shedding while the system is under abnormal conditions. Like dynamic routing in telecommunication networks, these power routers increase network reliability by decreasing and avoiding network congestion. Power routers with power flow control capability are required at important points in the main distribution network to meet the demand. A power router is a unit with power control capabilities and a "moderator", computer-based artificial intelligence software. The power flow controller (PFC) is a physical device comprised of a back-to-back converter that controls active and reactive power flow inside the feeders it is linked to. As shown in Figure 5.1, the PFC regulates the power flow through the feeders based on the arguments of the moderator [7, 8].

5.3   CYBER SECURITY-ENABLED POWER SYSTEMS

The power routing system must be very reliable and dynamic in order to build a highly reliable, scalable, and efficient power distribution network. Furthermore, the power router's real-time responsiveness is critical for maintaining an uninterrupted power supply. In addition, intelligent power routers should read information from other grid subsystems and work together with other power routers. As a result, in order to perform the implementation of this distribution network with power routing, a highly dependable and secret communication architecture is required. We suggest a cyber security enabled communication architecture as shown in Figure 5.2 that takes into account information confidentiality, integrity, and availability, as well as the smart grid's flexibility and scalability.

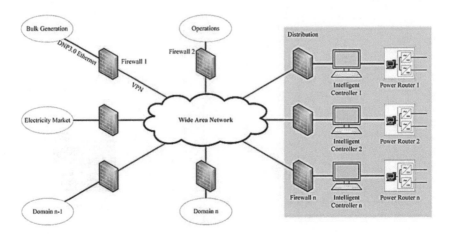

FIGURE 5.2 A cyber security-enabled communication architecture for power routing

A virtual private network (VPN) connection is utilized to exchange information between all the domains in the proposed architecture. If we select the most appropriate VPN framework with appropriate protocols for our application, the VPN can deliver a high level of security and real-time responsiveness. We chose the internet protocol security (IPSec) framework for our VPN because it can be customized in a variety of ways, has a variety of protocols for changing the levels of confidentiality, integrity, authentication, and secure key exchange, and operates at the network layer of the open systems interconnection (OSI) model [9–11].

The majority of unauthorized individuals who launch hacking assaults such as reconnaissance and access attacks do so from remote places. They may target our data as it travels over a wide area network. As a result, we must encrypt our data using a hashing method to ensure its confidentiality and integrity. The IPSec framework encrypts data packets with a pre-shared key and encryption method, generates a hash with a hashing technique, and encapsulates the data before sending them over the internet. As a result, anyone looking at our encrypted data from the internet will be unable to read it. Because the IPSec framework is unable to respond to social engineering attacks, staff must make extra efforts to thwart social engineering attackers. An authorized employee, for example, could be contacted by a hacker with an urgent problem that necessitates immediate network access. Furthermore, physical security is more critical than network and power system devices, because changes in the physical layer can cause all security protocols to fail.

REFERENCES

[1] J. Raiyn, "A survey of cyber attack detection strategies," *International Journal of Security and Its Applications*, vol. 8, no. 1, pp. 247–256, 2014, doi: 10.14257/ijsia.2014.8.1.23

[2] M. Uma and G. Padmavathi, "A survey on various cyber attacks and their classification," *International Journal of Network Security*, vol. 15, no. 5, pp. 390–396, 2013.

[3] T. H. Morris and W. Gao, "Industrial control system cyber attacks," *Proceedings of 1st International Symposium ICS & SCADA Cyber Security Research*, 2013, pp. 22–29, doi:10.14236/ewic/ICSCSR2013.3

[4] V. Zhou, C. Leckie, and S. Karunasekera, "A survey of coordinated attacks and collaborative intrusion detection," *Computers & Security*, vol. 29, no. 1, pp. 124–140, 2010, doi: 10.1016/j.cose.2009.06.008

[5] J. H. Jafarian, E. Al-Shaer, and Q. Duan, "An effective address mutation approach for disrupting reconnaissance attacks," *IEEE Transport Info Forensics and Security*, vol. 10, no. 12, pp. 2562–2577, 2015, doi: 10.1109/TIFS.2015.2467358

[6] U. Meyer and S. Wetzel, "A man-in-the-middle attack on UMTS," *Proceedings of 3rd ACM Workshop on Wireless Security*, 2004, pp. 90–97, https://doi.org/10.1145/1023646.1023662

[7] P. H. Nguyen, W. L. Kling, and P. F. Ribeiro, "Smart power router: A flexible agent-based converter interface in active distribution networks," *IEEE Transactions on Communications Smart Grid*, vol. 2, no. 3, pp. 487–495, 2011, doi: 10.1109/TSG.2011.2159405

[8] P. H. Nguyen, W. L. Kling, G. Georgiadis, M. Papatriantafilou, L. A. Tuan, and L. Bertling, "Distributed routing algorithms to manage power flow in agent-based active distribution network," *Innovative Smart Grid Technologies Conference Europe, 2010 IEEE Power and Energy Society*, 2010, pp. 1–7, doi: 10.1109/ISGTEUROPE.2010.5638951

[9] M. Z. Gunduz and R. Das, "Cyber-security on smart grid: Threats and potential solutions", *Computer Networks*, vol. 169, p. 107094, 2020. doi: 10.1016/j.comnet.2019.107094.

[10] A. Sani, D. Yuan, J. Jin, L. Gao, S. Yu and Z. Y. Dong, "Cyber security framework for Internet of Things-based energy internet," *Future Generation Computer Systems*, vol. 93, pp. 849–859, 2019. doi: 10.1016/j.future.2018.01.029.

[11] F. Al-Turjman and M. Abujubbeh, "IoT-enabled smart grid via SM: An overview", *Future Generation Computer Systems*, vol. 96, pp. 579–590, 2019. doi: 10.1016/j.future.2019.02.012.

6 Expert Systems for Microgrids

6.1 OPTIMIZATION OF ENERGY MANAGEMENT SYSTEMS FOR MICROGRIDS USING REINFORCEMENT LEARNING

6.1.1 SUPERVISED, UNSUPERVISED, AND REINFORCEMENT LEARNING

Supervised and unsupervised learning methods are the most commonly applied machine learning methods. In supervised learning, the learning model has access to labelled examples that guide its behavior. Unsupervised learning is about the representation and extraction of underlying structures in data. Recently, reinforcement learning (RL) has been introduced as an emerging machine learning approach because of its wide range of applications. Unlike the other learning approaches, the RL model generates its own training data by interacting with the environment, and learns the consequences of its actions through trial and error rather than by following predefined behavior patterns. The only evaluator for the RL agent is the "reward". The reward is capable of identifying and encouraging desirable behaviors, but cannot exactly guide the learner on how to achieve them. Table 6.1 presents a brief comparison between supervised and reinforcement learning approaches.

6.1.2 FUNDAMENTALS OF REINFORCEMENT LEARNING

6.1.2.1 General Reinforcement Learning Model

The learner in the RL model is called an "agent". This agent is a program that trains to achieve specific objectives. The platform in which the agent acts is called the "environment". The environmental condition at a specific time is called a "state". Specific moves that are taken by the agent in different states are called "actions". The agent's performance is evaluated by a scalar "reward" scheme. The specified reward promotes desirable actions by rewarding the agent. Figure 6.1 presents the general RL model.

The function of the RL model can be explained by considering a system transition from one state to another. When an agent is in a particular environmental state, it evaluates all the possible actions for that state. The selection of an action

DOI: 10.1201/9781003216292-6

TABLE 6.1

Comparison between Supervised and Reinforcement Learning Approaches

Supervised Learning	Reinforcement Learning
Learn patterns from the training data	Learn by trial and error sessions
Exploration of the environment is not necessary	Exploration of the environment is required
Expected answers are provided in training data	A policy should be developed to tell agents to take which action at which state
Generally follows a single decision process	Follows multiple decision processes

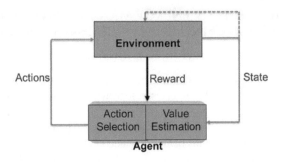

FIGURE 6.1 The general RL model

in a particular state is under an exploration and exploitation dilemma. However, the main goal of an RL agent is to maximize the rewards in the long run. Therefore, the action evaluation is based on the quality of its current state in terms of expected future rewards. The selection of the action with a maximum expected future reward is called the exploitation of the environment. Other than that, the agent can explore the environment by initiating random actions. Exploitation and exploration cannot be performed simultaneously by an agent. Therefore, the agent should decide when to explore and when to exploit, using action selection methods that will be discussed in an upcoming section. As an action is selected, the agent will get rewarded or penalized depending on its performance. This will result in a state transition. The process discussed above continues until the agent completes its learning phase.

6.1.2.2 Markov Decision Process

The Markov Decision Process (MDP) provides a general mathematical framework for the sequential decision making that defines the environment of RL. If a particular task can be formulated in the MDP, then the agent will have a platform where actions can be initiated while receiving rewards and experiences.

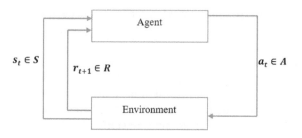

FIGURE 6.2 The MDP model

FIGURE 6.3 State transition in MDP

The MDP is based on the Markov property, which is a memoryless property. The Markov property says that the present state contains all the necessary information to predict the future. In addition to the current state, the MDP considers the selected action and the reward gained at the current state.

The MDP framework can be represented through a tuple $\langle S, A, P, R, \gamma \rangle$ where S is the set of states, A is the set of actions, P is the probability that action α in state s at time t will lead to state s' at time $t+1$ ($P(s,a,s')$), R is the immediate reward for the transition from s to s' ($R(s,a,s')$), and γ is the discount factor. The transition dynamics function (P) represents the stochastic nature of outcomes in an MDP environment, and therefore it is defined by a probability distribution. The general model of the MDP environment is as in Figure 6.2.

As shown in Figure 6.3, at time t the agent receives a state s_t from the set of all possible states (S). Depending on the received state s_t, the RL agent selects an action from the set of all possible actions for state s_t. In the next time step $t+1$, the agent receives a reward or a punishment depending on the action taken at the time step t. The agent's action in the state s_t results in a transition to s_{t+1}.

It is important to note that for the problem formation in MDP all five elements in $\langle S, A, P, R, \gamma \rangle$ should be known. However, in RL applications the transition dynamics function would not be expected to be known. This is a constraint for modeling. The RL environment modeling in MDP with partial information is not possible without sample-based learning algorithms such as Monte Carlo (MC) and temporal difference (TD) learning. Sample-based learning algorithms will be discussed in an upcoming section.

6.1.2.3 The Goal of the Reinforcement Learning Agent

The goal of an RL agent is to maximize the expected return (Gt). The expected return is basically the summation of all expected future rewards starting from the current state. Expected return formulae differ slightly depending on how the agents interact with the environment. There are two types of interaction, based on episodic and continuing tasks.

In episodic tasks, the interactions of the agent with the environment can be split into certain episodes with specified terminal states. The expected return from each episode can be determined simply by taking the summation of all possible future rewards starting from that state. The future expected reward at time t is the summation of possible rewards at time $t+1, t+2, ..., T$ as in equation (6.1).

$$G_t = R_{t+1} + R_{t+2} + R_{t+3} + \cdots + R_T \tag{6.1}$$

Continuing tasks are specified when an agent continuously interacts with the environment, without a specified terminal state. Since there is no terminal state, the future expected reward becomes infinite, as in equation (6.2). In order to make sure that Gt is finite, future rewards are discounted by a factor γ as given in equation (6.3) such that $0 \leq \gamma \leq 1$.

$$G_t = R_{t+1} + R_{t+2} + R_{t+3} + \cdots \tag{6.2}$$

$$G_t = R_{t+1} + \gamma R_{t+2} + \gamma^2 R_{t+3} + \cdots = \sum_0^\alpha \gamma^K R_{t+K+1} \tag{6.3}$$

If γ is set to 0, the agent becomes short-sighted, such that the agent only cares about the immediate reward. If γ is set to 1, the agent becomes a far-sighted agent that takes future rewards into account more strongly. In addition to that, future expected returns can be presented in a recursive form as derived in equation (6.4).

$$G_t = R_{t+1} + \gamma R_{t+2} + \gamma^2 R_{t+3} + \cdots$$

$$G_t = R_{t+1} + \gamma(R_{t+2} + \gamma R_{t+3} + \cdots)$$

$$G_t = R_{t+1} + \gamma G_{t+1} \tag{6.4}$$

6.1.2.4   Policies and Value Functions

For each state, there is a possible set of actions. An action should be selected out of the set of all possible actions. A policy states which action should be selected at each state. Therefore, a policy is basically a distribution over the possible actions for each state. There are two types of policies, deterministic and stochastic. Deterministic policies map each state to a single action. Stochastic policies select an action for each state out of multiple actions with non-zero probabilities.

As an agent selects actions according to a policy π, in each state the agent's performance is evaluated based on value functions. Value functions are crucial in reinforcement learning for an agent to identify the quality of its current state instead of waiting to observe the long-term outcome. This also paves the way for evaluating the quality of the current policy. There are two types of value functions, state value functions and action value functions. A state value function is the expected return from a given state, which is generally defined with respect to a policy π as in equation (6.5).

$$V_\pi \overset{\text{def}}{=} \mathrm{E}_\pi[G_t \mid S_t \qquad\qquad (6.5)$$

An action value function is the expected return from a given state if the agent selects an action following a policy π as in equation (6.6).

$$q_\pi(s,a) \overset{\text{def}}{=} \mathrm{E}_\pi[G_t \mid S_t \qquad\qquad (6.6)$$

Both state and action value functions can be derived as recursive formulae as follows. These recursive functions are called Bellman equations. Small MDPs can be directly solved through Bellman equations.

The optimal policy π^* is as good as or better than all the other policies, and it has the highest possible value function in all states. If the previously discussed action value function can be modified with respect to the optimal policy, optimal state and action value functions can be derived.

Bellman state and action value functions can be modified with the optimal policy to derive Bellman optimality equations. Here the policy π should be substituted by the optimal policy π^*.

However, this is not the conventional approach of RL. Generally, evaluating such specific policies is not the ultimate goal of RL. The main aim of RL is to learn an optimal policy that can maximize the cumulative reward in the long run. Optimal action value functions derived from Bellman optimality equations will be used in finding the optimal policy.

6.1.2.5 Sample-Based Learning

Sample-based learning methods are used to model the environment of the RL agent in MDP with partial information such that the transition dynamics of the MDP model remain unknown. These sample-based learning algorithms directly estimate value functions and policies from experience. Monte Carlo (MC) and temporal difference (TD) learning are two sample-based learning methods.

The MC method is a model-free learning method in which multiple returns are observed from a particular state. The observed returns are then averaged to estimate the expected return from that state. This is trial and error session-based learning, such that the agents in the system learn from the environment as they undergo interactive sessions and sample collection. Each "try" in an MC evaluation is called an "episode". As the MDP comes to its final stage, all the episodes should be terminated. The corresponding returns can only be observed at the end of an episode. The outcome of the task should be known before the learning can begin by evaluating the expected return for each state. Even though the MC approach allows the problem to be formed in the MDP framework, with partial information on the model, it still does not fully utilize the MDP structure. According to equation (6.7) for MC learning, the strategy updates take place after the completion of an entire episode, as the expected return term is stated in equation (6.4).

$$V\left(s_t\right) \leftarrow V\left(s_t\right) + \alpha\left[G_t - V\left(s_t\right)\right] \qquad (6.7)$$

Therefore, discussions have been held on making full use of the MDP structure in a more efficient manner. As a result, the temporal difference learning approach has been introduced. Unlike dynamic programming approaches, temporal difference learning is a model-free learning method where an environmental model is not required. As in the MC approach the agents learn directly from their experience. However, TD learning promotes online learning without waiting until the final outcome is available. According to equation (6.8), at each time step the TD estimations of the state value functions are updated. This an efficient use of the MDP structure.

$$V\left(s_t\right) \leftarrow V\left(s_t\right) + \alpha\left[R_{t+1} + \gamma V\left(s_{t+1}\right) - V\left(s_t\right)\right] \qquad (6.8)$$

There are on-policy and off-policy learning methods related to TD learning, which are known as SARSA and Q-learning, respectively.

6.1.2.6 On- and Off-Policy Learning Methods

In on-policy learning, the RL agent improves on and evaluates the policy that is being used to select actions. Unlike on-policy learning, off-policy learning improves on and evaluates a different policy from the one used for action selection. The policies that the RL agent learns and uses to choose actions are called the target policy and the behavioral policy, respectively. In on-policy learning methods such as SARSA, the target policy and the behavioral policy are the same, but for off-policy learning methods such as Q-learning, the target policy and the behavioral policy are different. Here, policy evaluation is the task of determining the state value function for a particular policy. Policy improvement is the task of improving the existing policy.

For example, let's consider the evaluation of a policy π_1 to obtain the state value V_{π_1}. Then we can greedify the policy with respect to V_{π_1} to obtain a better policy π_2. Likewise, we compute V_{π_2} to obtain an even better policy π_3. This is a sequence of better policies such that each policy is an improvement of the previous policy unless that is the optimal policy.

6.1.2.7 SARSA vs Q-Learning

SARSA is an on-policy learning method that always learns a near-optimal policy. The agent's environment exploration decays over time in SARSA. This learning method focuses on possible penalties from exploratory moves, such that the RL agent avoids optimal dangerous paths. Therefore, this learning method is suitable for training agents in the real world. SARSA is an action value form of TD learning. Here the agent only needs to know its next state–action pair, before performing updates on the action value estimations. SARSA alternatively performs policy evaluation and improvement.

Unlike SARSA, Q-learning directly learns the optimal policy while continuously exploring the environment at a fixed rate. This off-policy learning method focuses on triggering maximum rewards and neglecting possible penalties. Q-learning is suitable for training agents in simulation environments. The upcoming section explains the Q-learning function.

6.1.2.8 Q-Learning Algorithm

Q-learning has been introduced as the first and foremost online RL algorithm. As the name implies, each state has a value called a Q-value. As an agent visits a state, takes an action as an exploratory or exploitative move and gets rewarded, the corresponding Q-value will get updated. This assigned Q-value is an indication of the suitability of a particular action and is useful for the long run. These values are stored in a table called a Q-table.

In Q-learning, agents should thoroughly explore the MDP structure, for a considerably long period until the Q-values converge to Q*. Q-learning does not promote a dance of policy evaluation and improvement, such that the algorithm summarizes it to a single update. The general pseudo code for Q-learning is as follows. The Q-learning function can be presented using a flow chart as in Figure 6.4.

Initialize matrix Q to zero.
For each episode:
Select a random initial state.
Do while the goal state (convergent) hasn't been reached.
      Select one among all possible actions for the current state.
      Using this possible action, consider going to the next state.
      Get maximum Q-value for this next state based on all possible actions.
      Compute: $Q(s, a) \leftarrow Q(s, a) + \alpha [r + \gamma \max Q(s', a') - Q(s, a)]$
      $s \leftarrow s'$
      Set the next state as the current state.
End Do
End For

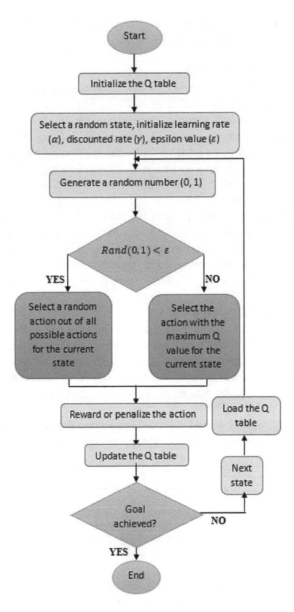

FIGURE 6.4   Q-learning algorithm

6.1.2.9   Exploration and Exploitation Strategy

Since RL agents set up their own dataset completely through the interaction with the environment, they experience a dilemma in the exploration and exploitation of the environment. Exploring the environment involves improving knowledge, which is beneficial in the long run. Exploiting the available knowledge, which is beneficial in maximizing immediate reward, is also an option. Since exploration and exploitation cannot be performed simultaneously, there are several strategies to determine when to explore and when to exploit, such as ε-greedy, upper confidence bound (UCB), SoftMax/Boltzmann, and random.

In ε-greedy action selection, a value called ε is determined before the learning begins. This ε is selected as a value between 0 and 1. According to ε-greedy action selection, the selected action will be the action with the maximum expected Q-value for that particular state with a probability of 1-ε and a random action with a probability of ε as in equation (6.9).

$$a_t = \begin{cases} argmax_a \ Q_t(a) \ with \ a \ probability \ 1- \in \\ uniform\{a_1,...,a_k\} \ with \ a \ probability \in \end{cases} \tag{6.9}$$

6.1.2.10 Hyperparameter Selection

The parameters of a model that are not manually set by the programmer are estimated or learned from a given dataset. In contrast, hyperparameters are externally defined values that cannot be determined from a given dataset. They are predefined to estimate the model parameters. In RL, there are several important hyperparameters, such as the learning rate, the decay rate, and the discount rate. The selection of the best combination of hyperparameters is called hyperparameter optimization or tuning. However, this can be treated as another problem by following different tuning algorithms. The most common approach is the trial and error method in which these parameters are tested using values assigned from the experience of the programmer. There are other structured methods, such as grid search and random search.

The importance of tuning can be highlighted if we consider an important hyperparameter in RL, the learning rate. Selection of the appropriate learning rate is important to overcome failures in training the model. In some cases, for a given learning rate the model will fail to reach an optimal policy or will require a longer duration to converge.

Let's make a comparison between SARSA and Q-learning approaches in the "Windy Grid World" environment. As shown in Figure 6.5, the Windy Grid World environment contains states called windy states. As an agent passes through these windy states, the wind flow will move the agent upwards. These windy states possess different wind strengths. An agent can move in four different directions. For each step taken, the agent will get a minus reward.

For this problem, both SARSA and Q-learning approaches are applied. Initially, the hyperparameters are $\varepsilon = 0.1$ and $\alpha = 0.5$. Figure 6.6 shows the simulation result for the variation of the number of completed episodes with time for both approaches.

When we analyze the performance of SARSA, it can be seen that the agent has performed slower learning at the beginning and then has improved in an exponential manner. Similarly, when we consider the performance of Q-learning for the same

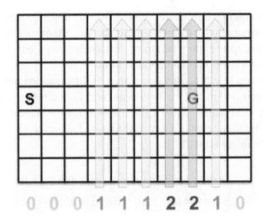

FIGURE 6.5 Windy Grid World environment

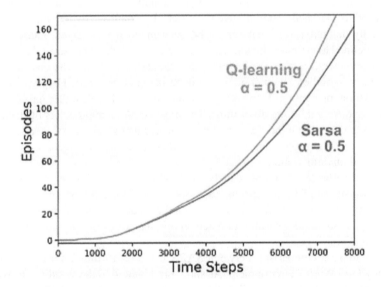

FIGURE 6.6 Variation in the number of completed episodes for SARSA and Q-learning approaches with $\varepsilon = 0.1$ and $\alpha = 0.5$

hyperparameters, it can be seen that the agent initially learned at a similar pace to the SARSA approach but then found a better policy.

An argument can be raised that SARSA considers its value function following ε-greedy behavior, while Q-learning focuses on the value function of the optimal policy, such that Q-learning performs faster learning. Before reaching a conclusion on the results obtained, let's reconsider the simulation with different learning rates for the two different approaches. Therefore, the SARSA algorithm will be modified

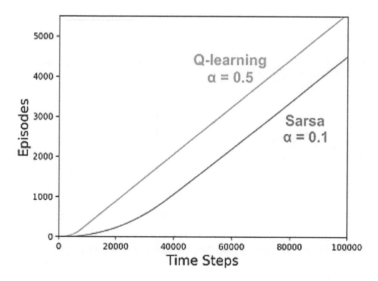

FIGURE 6.7 Variation in the number of completed episodes for SARSA ($\alpha = 0.1$) and Q-learning ($\alpha = 0.5$) approaches with $\varepsilon = 0.1$

with a new learning rate of 0.1 ($\alpha = 0.1$) and by continuing the experiment for a longer time than in the previous simulation.

According to the simulation result in Figure 6.7, both approaches have learned the same optimal policy as their lines are parallel towards the end, and also SARSA tends to learn slowly to perform as well as Q-learning. Therefore, we can conclude that hyperparameter optimization is important and varies with the simulation problem-solving approach even for a similar system. The learning rate, epsilon value and duration of the learning process can affect the final outcome.

6.1.3 SINGLE AND MULTI-AGENT REINFORCEMENT LEARNING

Single agent reinforcement learning (SARL) is associated with introducing a single learning agent to the system. Here, the intelligent agent individually carries out the learning process. Multi-agent reinforcement learning (MARL) involves more than one intelligent agent in the system. Here, the RL agents learn while cooperatively optimizing the system. They track the states in which agents cannot independently take actions and have to consider the decisions of fellow agents. They follow a strategy in which agents sort the states where collisions may occur and mark those states as "dangerous" states and other states as "safe" states. The general RL model for the multi-agent case is as in Figure 6.8.

Table 6.2 shows a comparison between SARL and MARL considering the features of their RL models. The main difference is that the multi-agent case comprises a joint action set where the actions of all the RL agents are recorded.

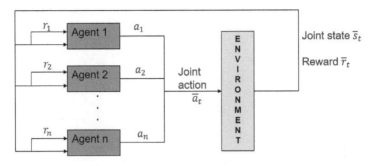

FIGURE 6.8   General MARL model

TABLE 6.2
Comparison between Single- and Multi-Agent RL Approaches

Single Learning Agent Case	Multiple Learning Agent Case
Defined by a Markov Decision Process. $\langle s, a, \rho \rangle$	Defined by a generalized Markov process, known as Stochastic Game. $\langle s, a_1, \ldots, \alpha_n, \rho_1, \ldots \rho_n \rangle$
$s \in S$ – A finite set of environment states	n – Number of agents, $i = 1, \ldots, n$
$a \in A$ – A finite set of agents' actions	$s \in S$ – A finite set of environment states
ρ – Reward function; $S \times A \times S \rightarrow\!\mid R$	$a_i \in A_i$ – A finite set of Agent i's actions
	$A = A_1 \times \ldots \times A_n$; A finite set of joint actions
	ρ_i; $S \times A \times S \rightarrow\!\mid R$ – Reward function of Agent i

In addition, the overall reward function comprises the reward functions of each individual agent.

6.1.4   Problem Formulation in RL

6.1.4.1   Defining the Goal

The objective of the application of RL should be clarified at the beginning. The special feature of RL is the ability to identify the optimal policy of operation purely through the interaction with the environment, without any labelled examples or correct answers. Therefore, this approach works for multiple decision process-based optimization problems in which objectives should be optimally achieved, rather than relying on a given dataset or manually defined algorithms. Figure 6.9 shows the process for determining the applicability of RL.

6.1.4.2   Mapping the Problem with RL Elements

In an RL system, there can be one or several learning agents. First, the environmental dynamics of each learning agent should be identified. These dynamics will form the

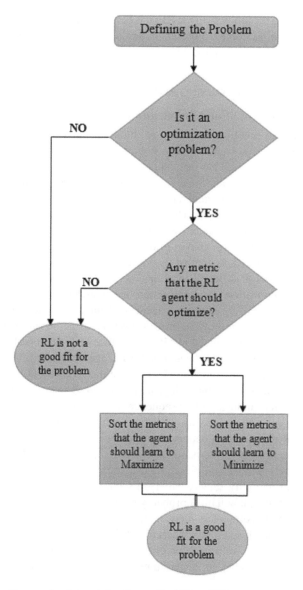

FIGURE 6.9   Process for determining the applicability of RL

states of the RL model. A state can be presented in the form of a tuple (such as (a, b, c…)). The next important task in modeling is to decide all the possible actions for each state. Then a set of actions will be available for each state, out of which the RL agent can select actions at each state following an exploiting move or an exploratory move. Defining a reward or a cost function is the most crucial step, since the reward/ cost function determines the effectiveness of the agent's learning process.

When mapping the problem to the RL, it is beneficial to study similar approaches to improve our model. Table 6.3 presents a summary of states, actions, and reward/cost functions defined for games, robot handling, and microgrid operations.

6.1.5 REINFORCEMENT LEARNING APPROACH FOR MICROGRIDS

6.1.5.1 Grid Consumption Minimization

RL introduces an intelligent nature to modern power systems, where the uncertainties introduced by the high penetration of renewables can be optimally managed. The RL approach has been applied in many studies for grid dependency minimization. In these studies, the main consideration is the optimum utilization of energy storage and renewable generation. This has been achieved using both SARL and MARL in different contexts. The most common approach is to introduce the intelligence to the local consumer, with a short-term objective of energy storage management and a long-term goal of grid consumption minimization.

As presented in [4], RL is applied for a grid-connected microgrid with two solar photovoltaic systems at different locations. Since the long-term objective is grid consumption minimization, local consumers are modeled as learning agents. The dynamic environment of the consumer agents is defined by solar power generation, battery level, and load variation. Conventional distributed optimization of the microgrid performance is compared with the MARL to evaluate the system performance. Here joint/cooperative Q-learning has been used for RL agent modeling.

A similar concept is presented in [5]. Here an n-step ahead Q-learning method is used for consumer agent modeling. n-step ahead Q-learning emphasizes the importance of determining an action sequence for a particular agent, even for future n+1 states. For example, let's consider battery scheduling for a solar microgrid in which only one action, out of battery charging and discharging, can be initiated at each state. At state s_t, let the demand be high and the solar generation be relatively low. If the battery should be discharged to maximize the immediate reward, the consumer agent naturally selects battery discharging. If the battery has already been drained at time t, it will most probably not be available at $t + 1$, $t + 2$, when the solar power may be completely unavailable. This raises the issue of not utilizing the solar power available at time t. Therefore, this study emphasizes the importance of widening the planning scope for grid consumption minimization. The same approach is followed in [6], for a wind power generation based microgrid.

6.1.5.2 Minimization of Demand–Supply Deficit

Demand–supply mismatch within a microgrid system should also be addressed. As described in [7], RL is applied to a group of microgrids with a common objective in a cooperative manner. Therefore, MARL is applied for these microgrids. MARL is implemented following cooperative Q-learning, in which microgrids share their battery status as common information and the Q-learning algorithm addresses the joint states and actions to be taken. This study aims to solve two energy management problems. The first objective is to minimize the demand–supply deficit through

TABLE 6.3
Problem Formation in RL for Different Applications

Task	Goal	RL Agent/s	States	Actions	Reward/Cost
Recycling robot for collecting empty soda cans [1]	Collecting the maximum possible number of cans within the shortest time	Robot	{Charge level of the robot}	If the charge level is low, {search, wait, recharge} If the charge level is high, {search, wait}	+100 reward for each successful can collection −1 reward for each energy unit consumed
Energy management of a solar microgrid [2]	Energy cost minimization of the microgrid	Consumer	{Grid price, net demand, energy stored}	Feeding the net demand with: Battery only, Grid only, Both battery and grid	Cost of energy purchased from the grid and the stored energy consumed
Market operation of a microgrid [3]	Maximizing the profit and finding the optimum bidding strategy for the agents	Renewable Energy source	{Time, main grid price}	Submitting the bid for energy selling	Profit of selling energy
		Diesel generator	{Time, main grid price}	Submitting the bid for energy selling	
		Storage	{Time, main grid price, SOC level}	Deciding whether to sell or buy and the amount to buy or sell Submitting the bid for energy selling	Profit of selling energy and the cost of buying energy

intelligent storage handling. The second objective is to minimize the cost of grid power consumed together with demand–supply balancing. The importance of this study is that it clearly models the objectives in terms of the reward functions, where the reward gets minimized as the demand–supply difference occurs. The results are improved through modification of the reward scheme.

In contrast to the AC power systems discussed above, [8] presents a grid-connected DC microgrid with solar and wind power sources, storage systems and loads. To define the problem in the RL framework, the DC microgrid components such as the DC bus current, voltage, and state of charge of the storages are mathematically modeled together with the constraints on power, power dynamics and state of charge, and an objective function to be minimized. The main aim is to manage the power flows in the DC microgrid, by taking decisions at each discrete time step for the controllable components in the system as well as the main grid. Here the storage systems act as controllable components. The state space for the problem is defined based on the storage devices, containing information related to state of charge, current, and relative demand in each discrete time step. The action space involves actions based on the current controlling for each storage device in the system. As in [7], the reward function also emphasizes minimizing the current flow from the grid, as the reward is a function of the main grid current flow.

6.1.5.3 Islanded Operation of Microgrids

RL has also been applied to microgrids when achieving autonomous operation in islanded mode. In [9], an islanded microgrid is considered with diesel units, battery banks, renewable sources, and critical and non-critical loads. Here, the suitability of applying Q-learning in problem-solving is evaluated. The study highlights that the conventional microgrid operation has a stochastic nature so the agents in the system may not know the upcoming states that result from their future actions. Also, for microgrid problem-solving, the expected Q-table may have large dimensions. Therefore, the Q-table updating process will take a large number of episodes. In order to address these challenges, [9] introduces a new approach for Q-learning by introducing a new variable called the state transition variable to represent the possible actions of agents and the transitions of the system. The advanced feature of this study is the absence of the grid. Therefore, for simulation purposes, it is replaced by a virtual slack bus, so that the main objective is to achieve zero power exchange with the slack.

6.1.5.4 Economic Dispatch

The economic dispatch problem relates to the optimization of the utilization of available generation sources following automatic generation control (AGC) commands in real time. In [10], a new, alternative RL approach is applied for the economic dispatch problem. A hierarchical correlated Q-learning algorithm, which promotes multi-layered optimal generation allocation, is the foundation of this research. Here, the RL agent for each generator optimizes its participation factor and cooperatively operates with other agents, taking their decisions into account. As previously discussed, in an alternative approach for Q-learning to handle the curse of dimensionality, the layered learning technique is used for this study.

In [11], a multi-agent system is introduced with the intention of modeling the customers, generators and storages as autonomous units through the RL approach. The main objective is to allow these agents to learn and develop their own energy management policies. This modeling is done with partial information, since in real scenarios consumers do not have complete information about battery scheduling and energy prices, and microgrids inherit a dynamic environment because of the uncertainties of renewable generation and demand. Here, five main intelligent agents, wind, PV, diesel, storage, and customer, are modeled by defining the expected actions and unique reward functions for each agent. This study highlights the effect of the RL approach on the system outcome by following four different configurations: without the involvement of learning agents, with intelligent distributed energy resources only (consumption is fixed), with intelligent consumers only, and with the involvement of all the agents as intelligent agents.

Similarly, ample research has been done on the application of RL for the economic dispatch problem [12].

6.1.5.5 Energy Market

Energy market operations are strongly related to energy cost estimations and the development of bidding strategies to maximize profits. Dynamic pricing in a microgrid has become more challenging because of the lack of prior information from the demand side and the intermittent nature of renewables. This is presented in [13], where dynamic pricing for a microgrid is done through a service provider. This service provider links customers with the power generation company. An RL approach is introduced, where model-free algorithms can be used for service providers and consumers to learn the optimal bidding strategy for the microgrid. Since off-policy, model-free algorithms such as Q-learning do not require a model of the system or environmental dynamics to initiate the learning procedure, this study emphasizes the advantage of RL for the market operation of a microgrid.

Similarly, [3] introduces an islanded microgrid, with a load shedding scheme. The proposed microgrid consists of four main agents: the distributed generator agent, the distributed storage agent, the consumer agent, and the controller agent. Power consumer agents have been modeled as intelligent agents using Q-learning, with a long-term goal of learning the optimal bidding strategy to maximize profit.

A summary of Q-learning applications on microgrids based on the objective is presented in Table 6.4.

6.2 CASE STUDY: REINFORCEMENT LEARNING APPROACH FOR MINIMIZING THE GRID DEPENDENCY OF A SOLAR MICROGRID

6.2.1 Proposed System

A grid-connected microgrid with a solar PV system is under consideration. This solar PV system consists of a solar array, a battery backup and a local consumer,

TABLE 6.4
Summary of Different RL Applications

Reference	System Type	RL Method	Exploration Strategy	Objective
[4]	PV, Battery, Loads, Grid	Joint Q-learning	N/A	Grid consumption reduction
[14]	PV, Battery, Loads, Grid	Coordinated Q-learning	N/A	
[7]	PV, Wind, Battery, Loads, Grid	Q-learning	e-greedy	Minimizing demand supply
[8]	PV, Wind, Storage (super capacitors and batteries), Loads, Grid	Q-learning	e-greedy/ SoftMax	deficit
[9]	Renewable source, Battery, Diesel, Critical and Non-critical loads	Q-learning	Boltzmann	Autonomous operation in the Islanded Mode
[15]	PV, Battery, Desalination unit, Local customer	Q-learning	N/A	
[5]	PV, Battery, Loads, Grid	3-step ahead Q-learning	N/A	Energy cost estimation
[2]	PV, Battery, Loads, Grid	Q-learning	e-greedy	
[6]	Wind, Battery, Loads, Grid	2-step ahead Q-learning	Deterministic	
[3]	PV, Wind, Micro turbines, Fuel cell, Loads, Grid	Q-learning	e-greedy	Energy market operation

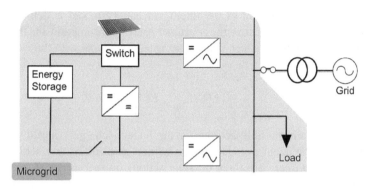

FIGURE 6.10   Proposed system

as shown in Figure 6.10. The long-term objective of applying RL is to minimize the grid dependency of the microgrid. This objective is to be achieved through the maximum utilization of available solar power generation and the energy stored. Four agents are assigned for the proposed system, the solar, battery, load and server agents, to introduce agent-based control.

The dynamics of the battery backup in the microgrid can be expressed as in equation (6.10). The battery energy level at a particular time is under a constraint as in equation (6.11).

$$E^{Battery}_{t+1} = E^{Battery}_t + E^{charge}_t - E^{discharge}_t \, (\text{Wh}) \tag{6.10}$$

$$E^{Battery}_{DoD} \leq E^{Battery}_t \leq E^{Battery}_{max} \tag{6.11}$$

According to equation (6.10), the battery energy level at a particular time step depends on the battery energy level and the charged or discharged energy amounts during the previous time step interval. Here E^{charge}_t and $E^{discharge}_t$ are the power flows, over the time step interval t to $t+1$, between the solar PV panel and the battery, and the battery and the consumer load, respectively. The energy storage unit is only charged by the connected solar PV panel. Also, the battery energy level at a particular time is constrained by a lower bound, which is the depth of discharge (DoD) level of the battery $(E^{Battery}_{DoD})$, and an upper bound, which is the maximum energy level of the battery ($E^{Battery}_{max}$).

6.2.2   SINGLE-AGENT REINFORCEMENT LEARNING MODEL

In SARL only the battery agent is modeled as an intelligent agent. These control agents are autonomous, social, reactive, and proactive in nature. In addition to the general characteristics of agents, learning ability is introduced through RL modeling.

The dynamic environment of the intelligent battery model is defined by the available solar power generation, the consumer energy consumption and the battery energy level. The state at time t can be presented by a tuple with three elements, as in equation (6.12).

$$S_t = \left(E_t^{PV}, E_t^{Load}, E_t^{Battery}\right) \qquad (6.12)$$

There are two main possible actions for the battery, charging and discharging the battery, and the agent can also remain without performing any action on the battery. Therefore, there are three possible actions on the energy storage, as in equation (6.13). The eligibility of the actions at a particular state depends on the battery energy level, such that it remains within the energy level limits as mentioned in the previous section. For example, if the battery energy level is at the lower bound, it will not be possible to execute the battery discharging action. These actions are under a constraint such that multiple actions cannot be executed simultaneously.

$$A_t = \left(a^{charge}, a^{discharge}, a^{idle}\right) \qquad (6.13)$$

The agent's performance evaluation function comprises two separate equations. Equation (6.14) presents the cost function for the realization of the long-term objective of the intelligent agent, which is to minimize the energy purchased from the grid to cater for the local demand. Here, the power flow into and from the grid (E_t^{grid}) is under consideration. Equation (6.15) is for the determination of E_t^{grid} for the different actions chosen. The terms m and n are the prices of purchasing a unit from the grid and selling a unit to the grid, respectively. The price of purchasing a unit from the grid follows a Time of Use (TOU) tariff structure.

$$\rho\left(s_t, a_t, s_{t+1}\right) = [\max\left(0, E_t^{grid}\right).m + \min\left(E_t^{grid}, 0\right).n] \qquad (6.14)$$

$$E_t^{grid} = \begin{cases} E_t^{Load} - (E_t^{PV} - E_t^{Battery})a = a^{charge} \\ E_t^{Load} - \left(E_t^{PV} + E_t^{Battery}\right)a = a^{discharge} \\ E_t^{Load} - E_t^{PV} a = a^{idle} \end{cases} \qquad (6.15)$$

Equation (6.16) presents a cost function to encourage the utilization of the generated solar energy and stored energy in the battery.

$$R\left(s_t, a_t, s_{t+1}\right) = \begin{cases} \min\left[E_t^{PV}, \left(E_{max}^{Battery} - E_t^{Battery}\right)\right]a = a^{charge} \\ \min\left[(E_t^{Load} - E_t^{PV}), \left(E_t^{Battery} - E_{DOD}^{Battery}\right)\right]a = a^{discharge} \\ 0 a = a^{idle} \end{cases} \qquad (6.16)$$

6.2.3 MULTI-AGENT REINFORCEMENT LEARNING MODEL

The proposed MARL model comprises two intelligent agents, namely the battery and the consumer agent.

In the MARL case, there are three approaches, based on the goals of the multiple intelligent agents: fully cooperative, fully competitive and mixed. For the proposed grid-connected microgrid, the mixed approach is under consideration, with neither intelligent agent being fully cooperative or fully competitive.

Here the RL model for the battery agent that was previously discussed under the SARL case is advanced by introducing an intelligent consumer agent. Maximizing the utility of the available solar energy and battery storage and maximizing the consumer comfort level are the short-term goals of the battery and consumer agents, respectively. Although they have different short-term goals, their long-term goal, to minimize the grid dependency, is common to them both.

The consumer agent follows a demand response-based approach to realize its short- and long-term goals. Generally, the total demand of the local consumers can be categorized into three types: fixed, meaning the demand that must be satisfied; power shiftable, meaning flexible demand such that the corresponding appliances can be set to operate in predefined power ranges; and time shiftable, meaning non-flexible demand that can only be shifted to off-peak periods. Here the proposed consumer agent is responsible for the power shiftable demand. There can be several such consumer agents dedicated to different types of appliances.

Figure 6.11 represents the RL model for the intelligent consumer agent. The dynamic environment of the consumer agent is formed from four variables, namely the solar energy available, the charge or discharge energy allowable, the power shiftable demand, and the total energy demand.

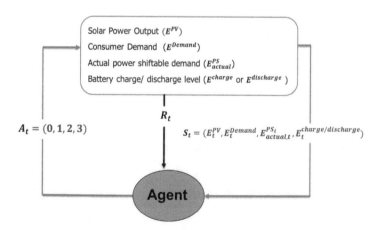

FIGURE 6.11 RL model for the intelligent consumer agent

The state tuple in equation (6.17) comprises these four variables that can be defined for each time interval.

$$s(t) = \left\langle E_t^{PV}, E_t^{demand}, E_{actual,t}^{PS_i}, E_t^{charge/discharge} \right\rangle \qquad (6.17)$$

At each state, the consumer agent selects an operating power range. These actions are denoted by discrete numbers, as in equation (6.18), which stand for different operating power ranges.

$$a(t) = \{0, 1, 2, 3\} \qquad (6.18)$$

The cost function of the consumer agent as in equation (6.19) denotes the level of satisfaction of the customer in realizing the requested demand at a particular time interval.

$$R_{PS_i}(t) = -\beta^{PS_i} \left(E_{actual,t}^{PS_i} - E_t^{PS_i} \right)^2 \qquad (6.19)$$

β^{PS_i} – dissatisfaction factor for power shiftable demand of consumer agent t

$E_{actual,t}^{PS_i}$ – actual power shiftable demand of consumer agent i at time t

$E_t^{PS_i}$ – supplied power shiftable demand of consumer agent i at titme t

The overall cost function of the multi-agent system is an extension of that of the SARL case, as presented in equation (6.20). Here the sum of the dissatisfaction costs of the total number of agents N is under consideration. Equations (6.21) and (6.22), respectively, are for the calculation of the grid energy flow and the new total demand of the system after the changes in power shiftable demand.

$$\rho(s_t, a_t, s_{t+1}) = [\max(0, E_t^{grid}).m + \min(E_t^{grid}, 0).n + \sum_{i=1}^{N} \beta^{PS_i} \left(E_{actual,t}^{PS_i} - E_t^{PS_i} \right)^2 \qquad (6.20)$$

$$E_t^{grid} = E_t^{demand_{new}} - E_t^{PV} - E_t^{discharge} + E_t^{charge} \qquad (6.21)$$

$$E_t^{demand_{new}} = E_t^{demand} - \sum_{i=1}^{N} \left(E_{actual,t}^{PS_i} - E_t^{PS_i} \right) \qquad (6.22)$$

6.2.4 SIMULATION MODEL

Figure 6.12 shows the simulation model for an intelligent agent in the system. Here the input data should be fed to the RL model to define the dynamic environment of the agent, and then the optimal operating policy can be extracted.

There are two approaches for feeding input data to the RL model: using past or statistical learning-based data. Statistical learning data are generated from machine learning-based prediction models such as artificial neural networks (ANNs).

6.2.4.1 Artificial Neural Network

An ANN is a network of neurons that process information. The ANN recreates the structure of the human brain. The basic model of a neural network is shown in

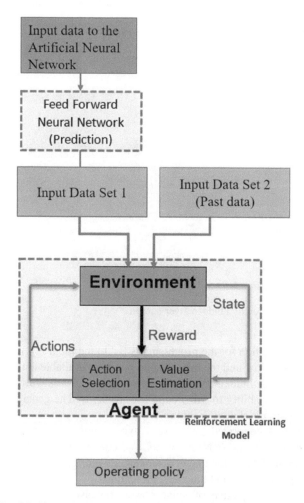

FIGURE 6.12 RL Simulation model for the RL agent

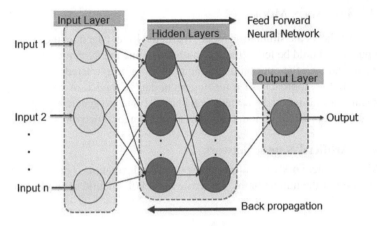

FIGURE 6.13   Structure of an artificial neural network

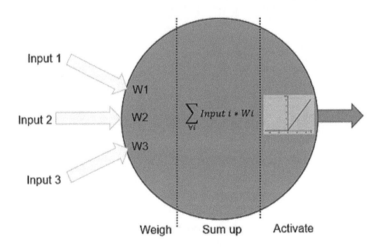

FIGURE 6.14   Function of a hidden layer neuron

Figure 6.13. Information for the processing is fed to the input layer. Hidden layers perform the processing, and the final desired output is available at the output layer. Models with multiple hidden layers are preferred over models with a single hidden layer based on their ability to solve complex problems. The general function of a hidden layer neuron is shown in Figure 6.14.

ANNs can be categorized into several types: multi-layer perceptron (MLP) networks, recurrent neural (RN) networks, and concurrent neural (CN) networks, based on the input data, hidden layer arrangement, etc. Table 6.5 gives a summary of the main types of ANN and their special features.

Considering the objective of solar power forecasting, an MLP network is selected. As the input to the MLP network, relevant weather data for the location

TABLE 6.5
Types of ANN

	Data Types	Special Features	Disadvantages

Multi-layer perceptron	Text data
	Tabular data
	Image data
		Learn non-linear functions
		Handle complex problems
		    that involve multiple input
		    parameters
			Vanishing gradient
			    problem
			Cannot capture
			    spatial data

Recurrent neural network	Time
	    series data
	Text data
	Audio data
		Capture sequential
		    information
		Parameter sharing
			relevant to image
			    processing

Concurrent neural network	Image data
	Video data
	Text data
		Capture spatial features
		Parameter sharing
		Learn filters autonomously to
		    capture relevant features
			Vanishing gradient
			    problem

of interest are used. The solar PV prediction using the MLP network generalizes the RL model by allowing further advancement of the system to a deep learning network. In addition, this is suitable for solar power generation prediction for a day or a year ahead for simulation purposes.

6.2.4.2 Feature Selection

Features are selected to identify the group of input parameters with the least root mean square error (RMSE). Figure 6.15 shows several RMSE plots for different input parameter groups. Considering the results, the parameter group with the least RMSE is selected for training. The proposed MLP network to feed the Q-learning algorithm is in Figure 6.16.

6.2.5 RL Simulation Models in Python

In this research, the simulations of the proposed SARL and MARL models are implemented through the Python platform. The default libraries such as Pandas, NumPy, TensorFlow, Matplotlib, etc. provide the framework for the initial RL model formation. There are separate Python programs for environment class, agent class, and data handling for the RL agent.

Environment Class: This class defines the state space environment of the intelligent agent and outputs the agent's operating state at each time interval. The available solar PV generation and the demand are fed to the environment class. The overall function of the environment class for the proposed system is given in Figure 6.17.

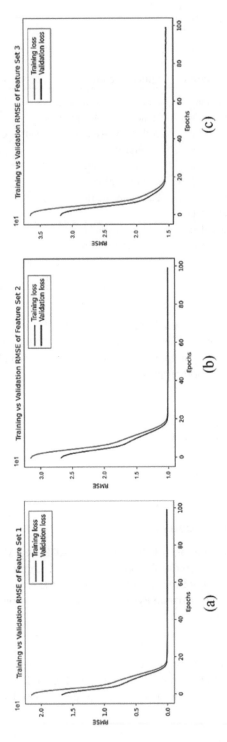

FIGURE 6.15 RMSE variation for different input parameter groups

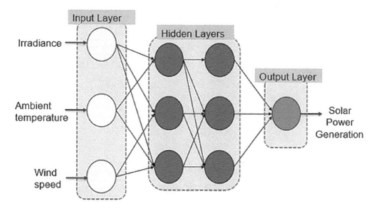

FIGURE 6.16 Proposed MLP network for solar PV prediction

Agent Class: This class contains the Q-learning algorithm. Here the agent selects an action by following the ε-greedy action selection method. Then, depending on the selected action, the deserved reward for the agent is determined. The agent class is responsible for initializing and loading the Q-table for all learning episodes. The overall function of the agent class for the proposed system is given in Figure 6.18.

Results

Table 6.6 presents a summary of the performances of the SARL and MARL models. The percentage reduction in grid consumption, the cost reduction and the improvement of battery utilization are analyzed as the performance indices.

The analysis of the simulation results shows that the SARL model is a better approach than the MARL model for system performance optimization. However, the proposed system can also be optimized to reach the expected outcomes through the MARL model by extending the learning period.

There are several reasons for these performance deviations. During the learning phase, the SARL agent only concentrates on balancing out knowledge improvement and exploitation, while the MARL agents additionally explore the behaviors of the other intelligent agents. This results in a dynamic best policy for the MARL model that makes the learning complicated because of stability issues. The involvement of multiple RL agents with different short-term goals, as in the proposed system, results in collisions between agents in satisfying their goals. Each intelligent agent in the proposed MARL model introduces a set of actions and states to the overall model that results in the exponential rise of variables. This case study concludes that the SARL model for the proposed grid-connected microgrid achieves the lowest cumulative grid energy consumption and the highest battery utilization index over the learning phase.

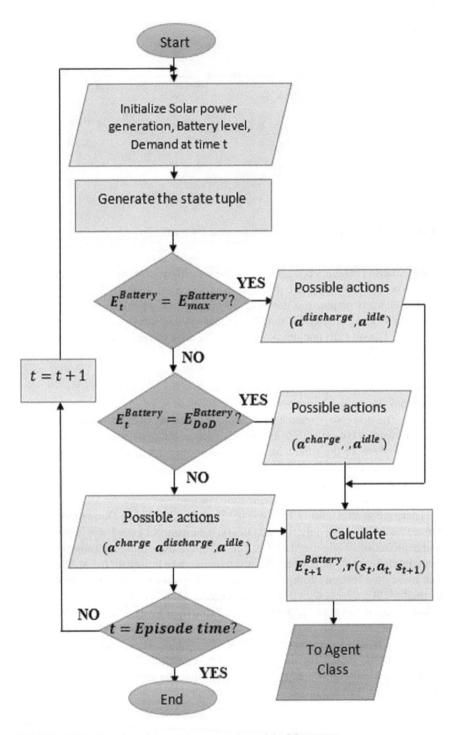

FIGURE 6.17   Function of the environment class of the RL agent

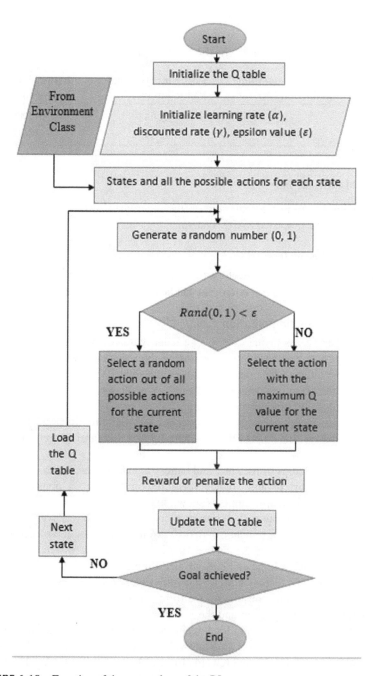

FIGURE 6.18   Function of the agent class of the RL agent

TABLE 6.6
Summary of the Performances of the SARL and MARL Models

Model	Grid Consumption Reduction (%)	Cost Function Value Reduction (%)	Battery Utilization Improvement (%)
SARL	27.027	33.333	71.429
MARL	18.919	15.625	58.333

6.2.6 HARDWARE IMPLEMENTATION

Here the hardware implementation of the microgrid testbed and agents are considered for the proposed SARL system.

6.2.6.1 Microgrid Testbed

A microgrid testbed can be developed for the high-performance SARL model. As shown in Figure 6.21, an AC microgrid can be developed as the proposed system.

6.2.6.2 Agent Implementation

There are four agents in the system: the solar, battery, load, and server agents. The solar agent is responsible for measuring the solar energy generation over a specified time interval. DC measurements of the solar PV current and voltage are recorded by the agent for the generated energy calculation.

The battery agent outputs the remaining battery energy level and calculates the charging or discharging energy level over a particular time interval. This agent actuates the relays that control battery charging and discharging according to the server's request.

The hardware setups of the solar and battery agents are the same. These agents comprise direct current and voltage measurement circuits, a power circuit, a microcontroller and a display unit, as shown in Figure 6.22.

The load agent measures the consumer energy consumption over a specified time interval. AC measurements of load current and voltage are recorded by the agent for the consumer demand calculation. The hardware setup of the load agent comprises AC voltage and current measurement circuits, a power circuit, a microcontroller and a display unit, as shown in Figure 6.23.

The server agent is responsible for the formulation of the dynamic environment of the intelligent battery agent and the RL algorithm. The recorded measurement data from the solar, battery and load agents are sent to the server agent to define the environmental state at each time interval. This generated state is fed to the RL algorithm to initiate actions and rewards for the Q-learning algorithm. The output of the server agent is the selected action for the battery backup, and is sent as a request to the battery agent to actuate the corresponding charging and discharging relays.

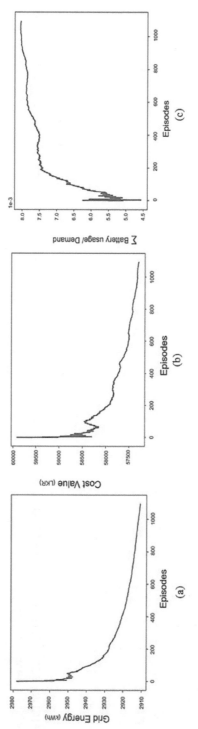

FIGURE 6.19 Simulation results for SARL model: (a) grid energy consumption variation; (b) cost value variation; and (c) battery usage variation with the learning episodes

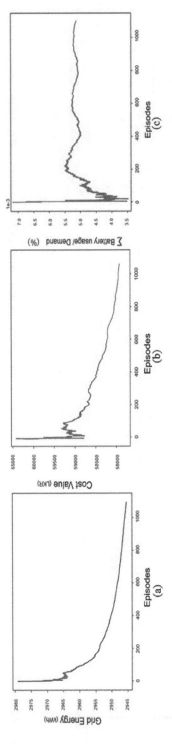

FIGURE 6.20 Simulation results for MARL model: (a) grid energy consumption variation; (b) cost value variation; and (c) battery usage variation with the learning episodes

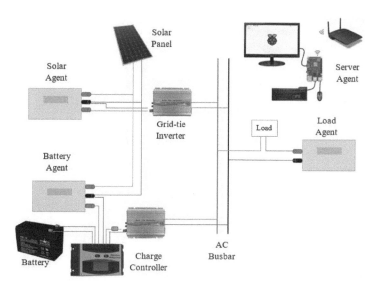

FIGURE 6.21 Microgrid testbed

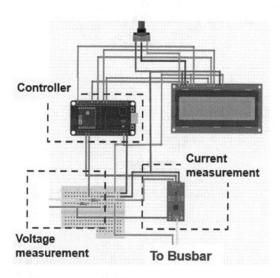

FIGURE 6.22 Hardware setup of solar and battery agents

The hardware setup of the server agent is a Raspberry pi setup as shown in Figure 6.24. Raspberry pi is a low-cost, small-scale computer that can be plugged into a monitor. It also supports external devices such as a standard keyboard, mouse, and ethernet connection. Like other microcontrollers, Raspberry pi also has the ability to input and output electric signals to actuate external electronic devices [16].

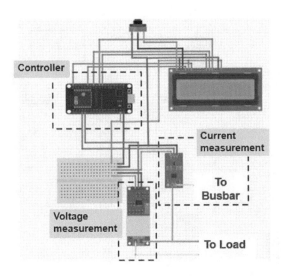

FIGURE 6.23   Hardware setup of load agent

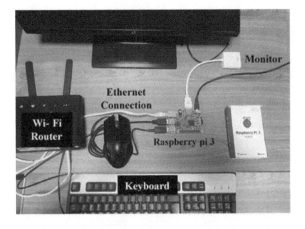

FIGURE 6.24   Raspberry pi hardware setup of server agent

In accordance with the proposed system, the Python-based RL algorithm can be run through a Raspberry pi setup. Figure 6.25 shows the Raspberry pi user interface. As shown in Figure 6.26 this controller supports Python programs.

6.2.7 Agent Communication

The controller selected for the agents is a Wi-Fi compatible microcontroller, namely ESP 32. Therefore, the agents communicate through Wi-Fi technology. The ESP 32 Wi-Fi module supports a distance range of about 100m with a data transfer range

FIGURE 6.25 Raspberry pi interface

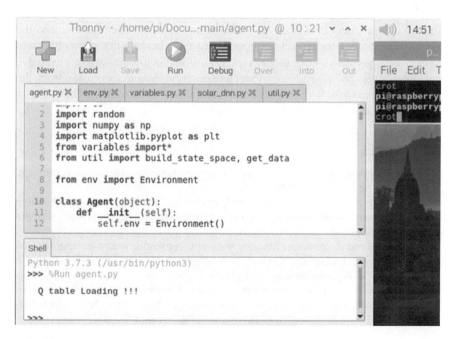

FIGURE 6.26 Python IDLE in Raspberry pi platform

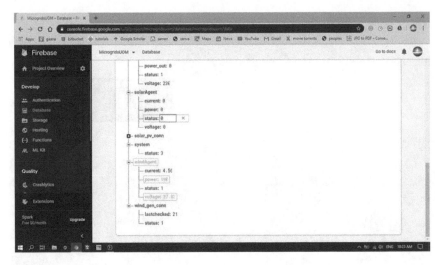

FIGURE 6.27   Firebase database interface

of 11–54 Mb/s. This controller also comes up with a desirable security level. The unit cost for the microcontroller is comparatively reasonable since no additional controlling modules are required for the operation [17].

6.2.8   FIREBASE DATABASE

Energy measurement data from the solar, battery and load agents are sent to the Firebase database, which is a cloud-hosted, real-time database that allows the users to store and sync data among themselves. Data stored in this database are available not only online but also offline. Therefore, energy measurement data from the agents can be used for both online and offline training of the system. The interface of the Firebase database is shown in Figure 6.27.

REFERENCES

[1]   R. S. Sutton and A. G. Barto, *Reinforcement learning: An introduction*. Cambridge, MA: The MIT Press, 2018.

[2]   S. Kim and H. Lim, "Reinforcement learning based energy management algorithm for smart energy buildings," *Energies*, vol. 11, no. 8, 2018, doi: 10.3390/en11082010.

[3]   Y. Lim and H. M. Kim, "Strategic bidding using reinforcement learning for load shedding in microgrids," *Comput. Electr. Eng.*, vol. 40, no. 5, pp. 1439–1446, 2014, doi: 10.1016/j.compeleceng.2013.12.013

[4]   L. Raju, R. S. Milton, and S. Sakthiyanandan, "Energy optimization of solar microgrid using multi-agent reinforcement learning," *Appl. Mech. Mater.*, vol. 787, pp. 843–847, 2015, doi: 10.4028/www.scientific.net/amm.787.843.

[5] R. Leo, R. S. Milton, and S. Sibi, "Reinforcement learning for optimal energy management of a solar microgrid," *2014 IEEE Glob. Humanit. Technol. Conf. – South Asia Satell. GHTC-SAS 2014*, pp. 183–188, 2014, doi: 10.1109/GHTC-SAS.2014.6967580.

[6] E. Kuznetsova, Y. F. Li, C. Ruiz, E. Zio, G. Ault, and K. Bell, "Reinforcement learning for microgrid energy management," *Energy*, vol. 59, pp. 133–146, 2013, doi: 10.1016/j.energy.2013.05.060.

[7] R. B. Diddigi, D. S. K. Reddy, and S. Bhatnagar, "Multi-agent Q-learning for minimizing demand-supply power deficit in microgrids," 2017, [Online]. Available: http://arxiv.org/abs/1708.07732.

[8] F. Lauri, G. Basso, J. Zhu, R. Roche, V. Hilaire, and A. Koukam, "Managing power flows in microgrids using multi-agent reinforcement learning," *Agent Technol. Energy Syst.*, pp. 1–8, 2013, http://citeseerx.ist.psu.edu/viewdoc/download?doi=10.1.1.700.1303&rep=rep1&type=pdf

[9] A. L. Dimeas and N. D. Hatziargyriou, "Multi-agent reinforcement learning for microgrids," *IEEE PES Gen. Meet. PES 2010*, pp. 1–8, 2010, doi: 10.1109/PES.2010.5589633.

[10] T. Yu, X. S. Zhang, B. Zhou, and K. W. Chan, "Hierarchical correlated Q-learning for multi-layer optimal generation command dispatch," *Int. J. Electr. Power Energy Syst.*, vol. 78, pp. 1–12, 2016, doi: 10.1016/j.ijepes.2015.11.057.

[11] E. Foruzan, L. K. Soh, and S. Asgarpoor, "Reinforcement learning approach for optimal distributed energy management in a microgrid," *IEEE Trans. Power Syst.*, vol. 33, no. 5, pp. 5749–5758, 2018, doi: 10.1109/TPWRS.2018.2823641.

[12] Y. Shang *et al.*, "Stochastic dispatch of energy storage in microgrids: An augmented reinforcement learning approach," *Appl. Energy*, vol. 261, p. 114423, 2020, doi: 10.1016/j.apenergy.2019.114423.

[13] B. Kim, Y. Zhang, M. Van Der Schaar, and J. Lee, "Dynamic pricing and energy consumption scheduling with reinforcement learning," *IEEE Trans. Smart Grid*, vol. 7, no. 5, pp. 2187–2198, 2016, doi: 10.1109/TSG.2015.2495145.

[14] L. Raju, S. Sankar, and R. S. Milton, "Distributed optimization of solar microgrid using multi-agent reinforcement learning," *Procedia Comput. Sci.*, vol. 46, pp. 231–239, 2015, doi: 10.1016/j.procs.2015.02.016

[15] P. Kofinas, G. Vouros, and A. I. Dounis, "Energy management in solar microgrid via reinforcement learning using fuzzy reward," *Advances in Building Energy Research*, vol. 12, no. 1, pp. 97–115, 2018, doi: 10.1080/17512549.2017.1314832.

[16] L. S. Melo, R. F. Sampaio, R. P. S. Leão, G. C. Barroso, and J. R. Bezerra, "Python-based multi-agent platform for application on power grids," *Int. Trans. Elec. Energy Syst.*, vol. 29, no. 6, 2019. doi: 10.1002/2050-7038.12012.

[17] L. Tightiz, H. Yang and M. Piran, "A survey on enhanced smart micro-grid management system with modern wireless technology contribution," *Energies*, vol. 13, no. 9, p. 2258, 2020. doi: 10.3390/en13092258.

7 Conclusion

Nowadays, fossil fuel-based electricity generation caters for a significant percentage of the total growing demand. Recently, people have become much more concerned about reducing fossil fuel-based electricity generation, as it contributes to environmental pollution through the emission of greenhouse gases, global warming, etc. which result in life-threatening conditions [1]. In addition, several international agreements on environmental protection have been introduced that allow countries to come up with environment protection protocols relevant to their contexts. For an example, the Paris agreement on climate change emphasizes the need for decarbonization to protect the planet from the impacts of global warming by keeping the global average temperature rise below 2°C by the end of the century [2].

The penetration of renewable energy sources such as solar, wind, etc. to the utility grid has therefore been identified as a timely solution to cater for the growing demand for electricity rather than being limited to large-scale, centralized, fossil fuel-based power plants [3, 4]. As a result, the distributed generation concept has come into the picture. Microgrids have been identified as a suitable integrating and controlling platform for these distributed energy resources (DERs) [5, 6]. Since DERs are mainly renewable resources, there are certain challenges in the high penetration of these sources due to the inherent issues with DERs such as intermittency and low reliability. Also, renewable sources are associated with power electronic interfaces that result in power losses due to power conversion stages, harmonics in voltage and current outputs, etc. [7, 8]. Therefore, a microgrid should have a continuous controlling and monitoring system to maintain the system functions, such as operating in multiple modes, economic dispatch, cost minimization, the optimum utilization of renewable sources, protection, etc.

However, the conventional control system provides centralized control, where a single failure can affect the overall operation of the power system. Microgrids prefer a distributed control system where complex control tasks are segregated into smaller tasks. In recent research interest, a novel distributed control method, multi-agent based controlling, has been introduced. Multi-agent systems (MAS) consist of multiple controlling agents that interact with each other to fulfil a complex task.

DOI: 10.1201/9781003216292-7

Microgrid fault detection, system isolation, and restoration of the system after fault clearance through the distributed controlling agents are presented in [9–12]. Recently, many studies have been done to evaluate MAS application for microgrid protection, where real-time monitoring of the system is essential [13–15]. In addition to the social, proactive, reactive, and autonomous qualities of agents, the aim is to introduce smart decision-making ability through learning. One or more intelligent agents can be introduced to the system to optimize the functions of the microgrid by learning optimal operational policies rather than relying on fixed algorithms [16–18]. Energy storage is another important component in modern microgrids, ensuring the demand–supply balance even in the islanded mode of operation when the grid is not available or in the absence of renewable generation [19, 20]. Battery units can also be controlled and optimized using optimization techniques to cater for peak demand, maximize utilization, minimize grid dependency, etc.

Microgrids operate in multiple modes such as grid-connected and islanded. Even though uni-directional power flows are present in the utility grid network, microgrids experience bi-directional power flows. Also, grid-feeding units become grid-forming units in the islanded mode. These multiple operating modes and the transitioning between modes require novel protection methods to withstand dramatically changing fault current levels, power flows, etc. Therefore, researchers are presently interested in developing adaptive protection for microgrids where the protection scheme changes with the detection of the operating mode [21–23]. Like these protection requirements, the structure of the required power electronic interfaces varies with the mode of operation. This has created a research gap for reconfigurable power systems. The development of reconfigurable inverters has been an advantage in terms of power loss reduction, quality improvement, and reliability [24–26].

However, there are security issues, such as illegal data acquisition, entries of external parties to confidential databases, service interruptions, etc. [27, 28]. Therefore, research has been done on enabling cyber security based power system operations through virtual private networks (VPNs) following an internet protocol security framework (IPsec) [29, 30].

Integrated microgrid platforms have become an emerging research area with the concept of networked microgrids. Most studies focus on a specific microgrid-related function such as developing adaptable protection systems, reconfigurable components, energy management systems, etc. Therefore, an overall solution for the microgrid functions through an integrated control platform still remains as a research gap. The design proposed in this book fulfils this research gap and, in addition, the interconnection of several such microgrids is also under consideration.

As the innovative features of this complete solution for microgrid controlling, intelligent energy storage units, reconfigurable power electronic interface units providing a combination of current, voltage and reactive power controlling, and communication infrastructures enabled to ensure cyber security can be highlighted.

REFERENCES

[1]   F. Martins, C. Felgueiras, M. Smitkova, and N. Caetano, "Analysis of fossil fuel energy consumption and environmental impacts in European countries," *Energies*, vol. 12, no. 6, p. 964 2019, doi: 10.3390/en12060964.

[2]   R. Warren, J. Price, J. VanDerWal, S. Cornelius, and H. Sohl, "The implications of the United Nations Paris Agreement on climate change for globally significant biodiversity areas," *Clim. Change*, vol. 147, no. 3–4, pp. 395–409, 2018, doi: 10.1007/s10584-018-2158-6.

[3]   M. F. Zia, E. Elbouchikhi, and M. Benbouzid, "Microgrids energy management systems: A critical review on methods, solutions, and prospects," *Appl. Energy*, vol. 222, pp. 1033–1055, 2018, doi: 10.1016/j.apenergy.2018.04.103.

[4]   I. Táczi, B. Hartmann, and I. Vokony, "Impact study of smart grid technologies on low voltage networks with high penetration of renewable generation," *Int. J. Renew. Energy Res.*, vol. 10, no. 2, pp. 519–528, 2020.

[5]   H. P. Pourbabak, T. C. Chen, B. Z. Zhang, and W. S. Su, "Control and energy management system in microgrids," in S. Obara and J. Morel (Eds.), Clean Energy Microgrids, pp. 109–133, 2017, doi: 10.1049/pbpo090e_ch3.

[6]   A. Hussain, V. H. Bui, and H. M. Kim, "Microgrids as a resilience resource and strategies used by microgrids for enhancing resilience," *Appl. Energy*, vol. 240, pp. 56–72, 2019, doi: 10.1016/j.apenergy.2019.02.055.

[7]   A. Marini, L. Piegari, S. S. Mortazavi, and M. S. Ghazizadeh, "Active power filter commitment for harmonic compensation in microgrids," *IECON Proc. (Industrial Electron. Conf.)*, pp. 7038–7044, 2019, doi: 10.1109/IECON.2019.8927668.

[8]   J. C. Wu, H. L. Jou, and X. Z. Wu, "Power conversion interface with harmonic suppression for a DC grid and single-phase utility," *IET Power Electron.*, vol. 13, no. 7, pp. 1302–1310, 2020, doi: 10.1049/iet-pel.2019.0630.

[9]   M. K. Perera *et al.*, "Multi agent based energy management system for microgrids," *PIICON 2020 – 9th IEEE Power India Int. Conf.*, pp. 6–11, 2020, doi: 10.1109/PIICON49524.2020.9113021.

[10] H. F. Habib and O. Mohammed, "Decentralized multi-agent system for protection and the power restoration process in microgrids," *IEEE Green Technol. Conf.*, pp. 358–364, 2017, doi: 10.1109/GreenTech.2017.58.

[11] L. Raju, A. A. Morais, R. Rathnakumar, S. Ponnivalavan, and L. D. Thavam, "Micro-grid grid outage management using multi-agent systems," *2017 2nd Int. Conf. Recent Trends Challenges Comp. Models (ICRTCCM)*, pp. 363–368, 2017, doi: 10.1109/ICRTCCM.2017.21.

[12] H. V. V. Priyadarshana, M. A. K. Sandaru, K. T. M. U. Hemapala, and W. D. A. S. Wijayapala, "A review on multi-agent system based energy management systems for micro grids," *AIMS Energy*, vol. 7, no. 6, pp. 924–943, 2019, doi: 10.3934/ENERGY.2019.6.924.

[13] E. Abbaspour, B. Fani, and E. Heydarian-Forushani, "A bi-level multi agent-based protection scheme for distribution networks with distributed generation," *Int. J. Electr. Power Energy Syst.*, vol. 112, pp. 209–220, 2019, doi: 10.1016/j.ijepes.2019.05.001.

[14] M. S. Rahman, T. F. Orchi, S. Saha, and M. E. Haque, "Multi-agent approach for overcurrent protection coordination in low voltage microgrids," *IEEE Power Energy Soc. Gen. Meet.*, 2019, doi: 10.1109/PESGM40551.2019.8974053.

[15] H. F. Habib, T. Youssef, M. H. Cintuglu, and O. A. Mohammed, "Multi-agent-based technique for fault location, isolation, and service restoration," *IEEE Trans. Ind. Appl.*, vol. 53, no. 3, pp. 1841–1851, 2017, doi: 10.1109/TIA.2017.2671427.

[16] X. Lu, X. Xiao, L. Xiao, C. Dai, M. Peng, and H. V. Poor, "Reinforcement learning-based microgrid energy trading with a reduced power plant schedule," *IEEE Internet Things J.*, vol. 6, no. 6, pp. 10728–10737, 2019, doi: 10.1109/JIOT.2019.2941498.

[17] T. Levent, P. Preux, E. Le Pennec, J. Badosa, G. Henri, and Y. Bonnassieux, "Energy management for microgrids: A reinforcement learning approach," *Proc. 2019 IEEE PES Innov. Smart Grid Technol. Eur. ISGT-Europe 2019*, 2019, doi: 10.1109/ISGTEurope.2019.8905538.

[18] W. Liu, P. Zhuang, H. Liang, J. Peng, and Z. Huang, "Distributed economic dispatch in microgrids based on cooperative reinforcement learning," *IEEE Trans. Neural Networks Learn. Syst.*, vol. 29, no. 6, pp. 2192–2203, 2018, doi: 10.1109/TNNLS.2018.2801880.

[19] M. Tavakkoli, E. Pouresmaeil, R. Godina, I. Vechiu, and J. P. S. Catalão, "Optimal management of an energy storage unit in a PV-based microgrid integrating uncertainty and risk," *Appl. Sci.*, vol. 9, no. 1, p. 169, 2019, doi: 10.3390/app9010169.

[20] X. Li and S. Wang, "A review on energy management, operation control and application methods for grid battery energy storage systems," *CSEE J. Power Energy Syst.*, vol. 7, no. 5, pp. 1026–1040, 2019, doi: 10.17775/cseejpes.2019.00160.

[21] A. A. Khazaei and A. Mahmoudi, "Decentralized adaptive protection structure for microgrids based on multi-agent systems," *2019 Iran. Conf. Renew. Energy Distrib. Gener. ICREDG 2019*, 2019, doi: 10.1109/ICREDG47187.2019.194143.

[22] O. Núñez-Mata, R. Palma-Behnke, F. Valencia, P. Mendoza-Araya, and G. Jiménez-Estévez, "Adaptive protection system for microgrids based on a robust optimization strategy," *Energies*, vol. 11, no. 2, p. 308, 2018, doi: 10.3390/en11020308.

[23] W. L. T. Peiris, W. H. Eranga, K. T. M. U. Hemapala, and W. D. Prasad, "An adaptive protection scheme for small scale microgrids based on fault current level," *2018 2nd Int. Conf. Electr. Eng. EECon 2018*, pp. 64–70, 2018, doi: 10.1109/EECon.2018.8540992.

[24] A. Kavousi-Fard, A. Zare, and A. Khodaei, "Effective dynamic scheduling of reconfigurable microgrids," *IEEE Trans. Power Syst.*, vol. 33, no. 5, pp. 5519–5530, 2018, doi: 10.1109/TPWRS.2018.2819942.

[25] R. K. Singh, A. Kumar, V. Kumar, A. Kumar, and P. M. Chavan, "Single stage single phase reconfigurable inverter topology," *Int. Res. J. Eng. Tech.*, vol. 6, no. 5, pp. 2352–2356, 2019.

[26] K. A. H. Lakshika, M. K. Perera, W. D. Prasad, K. T. M. U. Hemapala, V. Saravanan and M. Arumugam, "Z-source inverter based reconfigurable architecture for solar photovoltaic microgrid," *2020 IEEE Region 10 Symposium (TENSYMP)*, 2020, pp. 1543–1546, doi: 10.1109/TENSYMP50017.2020.9230895.

[27] A. Dagoumas, "Assessing the impact of cybersecurity attacks on power systems," *Energies*, vol. 12, no. 4, p. 725, 2019, doi: 10.3390/en12040725.

[28] X. Liu, Z. Li, Z. Shuai and Y. Wen, "Cyber attacks against the economic operation of power systems: A fast solution," *IEEE Trans. Smart Grid*, vol. 8, no. 2, pp. 1023–1025, 2017, doi: 10.1109/TSG.2016.2623983.

[29] W. H. Eranga, W. L. T. Peiris, M. K. Perera, and K. T. M. U. Hemapala, "Cyber-security enabled communication architecture for power routing in the smart grid," *2020 IEEE Int. Conf. Computing, Power and Communication Technologies (GUCON)*, 2020, pp. 729–733, doi: 10.1109/GUCON48875.2020.9231125.

[30] G. N. Ericsson, "Cyber security and power system communication – Essential parts of a smart grid infrastructure," *IEEE Trans. Power Deliv.*, vol. 25, no. 3, pp. 1501–1507, 2010, doi: 10.1109/TPWRD.2010.2046654.

Index

Note: Page numbers in **bold** refer to tables; page numbers in *italic* refer to figures.

A

Adaptive protection, 79, 81–83, 86, 94, 140, 142
  general algorithms for designing, 82
  implementation of, 83
Agent communication, 89, 91
Artificial neural network, 123–124

C

Cascading Failure, 2–3
Control architecture, 63, 74–75
  centralized, 40
  decentralized, 63
  distributed, 63
  types of microgrid, 40
Conventional power grid, 1–2
Converters
  AC/AC, 34
  AC/DC, 34
  boost, 67
  buck, 67
  buck–boost, 67
  DC to DC, 34, 65–67
  DC/AC, 34
  DC/AC/DC, 34
  Voltage source, 43
Cyber attacks, 96, 143
  Industrial control system, 99
  overview of, 95
  survey on various, 99
Cyber security, 95, 97–99, 140
  communication infrastructures enabled to
    ensure, 140
  SCADA, 99

D

Demand–supply deficit, 37, 114
Distributed generation, 81, 93, 139, 141
  advantages of, 9–10
  decentralized power system with more, 4
  dependable and cost-effective platforms for, 81
  through proper regulation and conversion, 9
Distributed generators, 39, 45
  connected to the grid as, 13
  dispatchable, 38
  grid-following, 39

non-dispatchable, 38
types of, 38

E

Electrical load, 39–41
Electrochemical battery, 31
Energy management, 33, 36, 44, 63, 88, 94, 101,
    114, **115**, 117, 136–137, 140–142
  microgrid, 45, 63, 87, 137, 141
  optimal distributed, 137
  tertiary control involves, 42
Energy market, 117–118
Energy storage systems, 30, 46, 116

F

Firebase database, 136
Flywheel
  benefits of, 32
  general structure of, 31
  kinetic energy in, 31
  round-trip conversion efficiency of, 31

H

Hyperparameter selection, 109

J

JAVA Agent Development Environment, 89

M

Markov Decision Process, 102–112
Microgrid protection, 81
  create novel adaptive, 86
  evaluate MAS application for, 140
Microgrid testbed, 130–133
Microgrids, 33–34, 42–43, 45–46, 63–64, 74, 81,
    94, 114, 116–117, 136–137, 139–142
  advantages of, 45
  approach for resilience improvement for, 46
  disadvantages of, 45
  energy management methods for, 33
  grid-connected, 36
  integrating distributed energy resources to, 65
  islanded operating mode of, 43
  modeling of solar, 75

networking of, 45–46
overview of, 33
proactive management of, 46
several limitations related to, 45
Modes of operation
  adequate protection in both, 86
  fluctuations in fault current levels during
    two, 81
  seamless transition between multiple, 33
Multi-agent, 86, 94, 111–112, 117, 122, 137,
    139, 141–142
  energy optimization of solar microgrid
    using, 136
  general RL model for, 111
  interaction between agents, 87
  python based, 137

P

Power electronic interfaces, 34, 65, 139
  important functions of, 65
  structure of required, 140
Power routing concept, 96

Q

Q-Learning, 106–107, 109–11, 114, 116, 127,
    130, 137
  actions and rewards for, 130
  alternative approach for, 116
  cooperative, 114
  coordinated, **118**
  general pseudo code for, 107
  hierarchical correlated 116, 137
  MLP network to feed, 125
  model-free algorithms such as, 117
  n-step ahead, 114
  off-policy learning methods such as, 106

R

Reinforcement learning, 101, **102**, 105, 117,
    136–137, 139
  A. G. Barto, 136
  dynamic pricing and energy consumption
    scheduling with, 137
  energy management in solar microgrid
    via, 137
  energy management systems for microgrids
    using, 101

fundamentals of, 101
the goal of, 103
multi agent, 111, 121, 137
R. S. Sutton, 136
single agent, 111, 119
strategic bidding using, 136
Renewable energy systems, 10–11
Renewable generation, 4, 11, 45, 114,
    140–141
  maximizing the utilization of, 33
  uncertainties of, 117
Renewable Generation Technologies, 11

S

Sample-based learning
  MDP with partial information, 103
  Monte Carlo (MC) and temporal difference
    (TD) learning, 105
SARSA, 106–107, 109–111
  environment exploration, 107
  off-policy learning methods related to TD
    learning, 106
  performance of, 109
Simulation models, 125
Smart grid, 4–8, 63–64, 93, 99, 137, 141–143
Smart power systems, 1
Solar energy, 11–12, 30, 46, 121
  maximizing the utility of the available, 121
  measuring, 130
  science behind, 12
  utilization of generated, 120
Supervised learning, 101

U

Unsupervised learning, 101

W

wind energy, 11, 46
  basics of, 25
  different configurations of, 27
  synchronization process of, 29

Z

Z source inverter, 142
  modeling of solar microgrids with, 75
  structure of, *74*